PRACTICAL ASPECTS OF
FINITE ELEMENT MODELLING
OF POLYMER PROCESSING

PRACTICAL ASPECTS OF FINITE ELEMENT MODELLING OF POLYMER PROCESSING

Vahid Nassehi

Chemical Engineering Dept., Loughborough University

JOHN WILEY & SONS, LTD

Chemistry Library

Copyright © 2002 by John Wiley & Sons, Ltd
         Baffins Lane, Chichester,
         West Sussex PO19 1UD, England

         National   01243 779777
         International (+44) 1243 779777

e-mail (for orders and customer service enquiries): cs-books@wiley.co.uk
Visit our Home Page on http://www.wiley.co.uk
                           or
                 http://www.wiley.com

All Rights Reserved. No part of this publication may be reproduced, stored in a retrieval
system, or transmitted, in any form or by any means, electronic, mechanical,
photocopying, recording, scanning or otherwise, except under the terms of the
Copyright, Designs and Patents Act 1988 or under the terms of a licence issued by the
Copyright Licensing Agency Ltd, 90 Tottenham Court Road, London W1P 0LP, UK,
without the permission in writing of the publisher.

Other Wiley Editorial Offices

John Wiley & Sons, Inc., 605 Third Avenue,
New York, NY 10158-0012, USA

Wiley-VCH Verlag GmbH, Pappelallee 3,
D-69469 Weinheim, Germany

John Wiley & Sons Australia Ltd, 33 Park Road, Milton,
Queensland 4064, Australia

John Wiley & Sons (Asia) Pte Ltd, 2 Clementi Loop #02-01,
Jin Xing Distripark, Singapore 129809

John Wiley & Sons (Canada) Ltd, 22 Worcester Road,
Rexdale, Ontario M9W 1L1, Canada

Library of Congress Cataloging-in-Publication Data

Nassehi, Vahid.
    Practical aspects of finite element modelling of polymer processing / Vahid Nassehi.
       p. cm.
    Includes bibliographical references.
    ISBN 0-471-49042-3
    1. Polymers—Mathematical models. 2. Chemical processes—Mathematical models.
    3. Finite element method. I. Title.

    TP1120 .N37 2001
    668.9 – dc21

                                                                          2001045560

British Library Cataloguing in Publication Data

A catalogue record for this book is available from the British Library

ISBN 0 471 49042 3   αc✻

Typeset in 10½/12½pt Times by Mayhew Typesetting, Rhayader, Powys
Printed and bound in Great Britain by Antony Rowe, Chippenham, Wiltshire
This book is printed on acid-free paper responsibly manufactured from sustainable
forestry, in which at least two trees are planted for each one used for paper production.

T P
1120
N37
2002
CHEM

To Kereshmeh

Contents

Preface

Computational fluid dynamics is a major investigative tool in the design and analysis of complex flow processes encountered in modern industrial operations. At the core of every computational analysis is a numerical method that determines its accuracy, reliability, speed and cost effectiveness. The finite element method, originally developed by structural engineers for the numerical modelling of solid mechanical problems, has been established as a powerful technique that provides these requirements in the solution of fluid flow and heat transfer problems. The most significant characteristic of this technique is its geometrical flexibility. Therefore, it is regarded as the method of choice in the analysis of problems posed in geometrically complex domains. For this reason the analysis of industrial polymer processing flow regimes is often based on the finite element technique.

Industrial polymer processing encompasses a wide range of operations such as extrusion, coating, mixing, moulding, etc. for a multiplicity of materials carried out under various operating conditions. The design and organization of each process should therefore be based on a detailed quantitative analysis of its specific features and conditions. The common – and probably the most important – part in the majority of these analyses is, however, the simulation of a non-isothermal, non-Newtonian fluid deformation and flow process.

Computer simulation of non-isothermal, non-Newtonian flow processes starts with the formulation of a mathematical model consisting of the governing equations, arising from the laws of conservation of mass, energy, momentum and rheology which describe the constitutive behaviour of the fluid, together with a set of appropriate boundary conditions. The formulated mathematical model is then solved via a computer based numerical technique. Therefore, the development of computer models for non-Newtonian flow regimes in polymer processing is a multi-disciplinary task in which numerical analysis, computer programming, fluid mechanics and rheology each form an important part. It is evident that these subjects cannot be covered in a single text, and an in-depth description of each area requires separate volumes. It is not, however, realistic to assume that before embarking on a project in the area of computer modelling of polymer processing one should acquire a thorough theoretical knowledge in all of these subjects. Indeed the normal time period allowed for completion of postgraduate research studies or industrial projects precludes such an ambitious requirement.

The utilization of commercially available finite element packages in the simulation of routine operations in industrial polymer processing is well established. However, these packages cannot be usually used as general research tools. Thus flexible 'in-house'-created programs are needed to carry out the analysis required in the investigation, design and development of novel equipment and operations.

This book is directed towards postgraduate students and practising engineers who wish to develop finite element codes for non-Newtonian flow problems arising in polymer processing operations. The main goal has been to enable the reader to come to speed in a relatively short time. Inevitably, in-depth discussions about the fundamental aspects of non-Newtonian fluid mechanics and mathematical background of the finite element method have been avoided. Instead, the focus of the text is to provide the 'parts and tools' required for assembling finite element models which have applicability in situations expected to arise under realistic conditions. The illustrative examples that are included in the book have been selected carefully to give a wide-ranging view of the application of the described finite element schemes to industrial problems.

The finite element program listed in the last chapter can be used to model non-isothermal steady-state generalized Newtonian flow in two-dimensional planar domains. The code solves laminar incompressible Navier–Stokes equations for a power law fluid. The program is written using a clear and simple style and does not include any special features and hence can be readily compiled using most Fortran compilers. The basic code given in this program may be extended to solve more complex polymer processing problems using the finite element schemes derived in the book (Chapter 4). An illustrative example that shows the extension of the code to axisymmetric flow problems is discussed in the text and the required modifications are highlighted on the program listing.

Vahid Nassehi

1

The Basic Equations of Non-Newtonian Fluid Mechanics

Computer modelling provides powerful and convenient tools for the quantitative analysis of fluid dynamics and heat transfer in non-Newtonian polymer flow systems. Therefore these techniques are routinely used in the modern polymer industry to design and develop better and more efficient process equipment and operations. The main steps in the development of a computer model for a physical process, such as the flow and deformation of polymeric materials, can be summarized as:

- formulation of a set of governing equations which in conjunction with appropriate initial and boundary conditions provide a mathematical model for the process, and

- solution of the formulated model by a computer based-numerical scheme.

Industrial scale polymer forming operations are usually based on the combination of various types of individual processes. Therefore in the computer-aided design of these operations a section-by-section approach can be adopted, in which each section of a larger process is modelled separately. An important requirement in this approach is the imposition of realistic boundary conditions at the limits of the sub-sections of a complicated process. The division of a complex operation into simpler sections should therefore be based on a systematic procedure that can provide the necessary boundary conditions at the limits of its sub-processes. A rational method for the identification of the sub-processes of common types of polymer forming operations is described by Tadmor and Gogos (1979).

Non-Newtonian flow processes play a key role in many types of polymer engineering operations. Hence, formulation of mathematical models for these processes can be based on the equations of non-Newtonian fluid mechanics. The general equations of non-Newtonian fluid mechanics provide expressions in terms of velocity, pressure, stress, rate of strain and temperature in a flow domain. These equations are derived on the basis of physical laws and rheological experiments. Because of the predominant role of non-Newtonian flow equations in the modelling of polymer processes it is important to understand the theoretical foundations of these equations. However, detailed explanation of the theoretical foundations of non-Newtonian fluid mechanics is outside the scope of the present book. The subject is covered in many textbooks devoted to the topic. For example, the reader can find detailed derivations of the basic equations of non-Newtonian fluid mechanics in Bird *et al.* (1960) and Aris (1989) and more specifically for polymeric fluids in Middleman (1977) and Bird *et al.* (1977).

In this chapter the general equations of laminar, non-Newtonian, non-isothermal, incompressible flow, commonly used to model polymer processing operations, are presented. Throughout this chapter, for the simplicity of presentation, vector notations are used and all of the equations are given in a fixed (stationary or Eulerian) coordinate system.

1.1   GOVERNING EQUATIONS OF NON-NEWTONIAN FLUID MECHANICS

1.1.1   Continuity equation

The continuity equation is the expression of the law of conservation of mass. This equation is written as

$$\nabla.v = 0 \tag{1.1}$$

where ∇ is the operator nabla (gradient operator) and v is the velocity vector. Equation (1.1) is also called the incompressibility constraint. The absence of a pressure term in the above equation is a source of difficulty in the numerical simulation of incompressible flows.

1.1.2   Equation of motion

The equation of motion is based on the law of conservation of momentum (Newton's second law of motion). This equation is written as

$$\rho \frac{\partial v}{\partial t} + \rho v \cdot \nabla v = \nabla \cdot \boldsymbol{\sigma} + \rho \boldsymbol{g} \tag{1.2}$$

where ρ is fluid density, v is velocity, $\boldsymbol{\sigma}$ is the Cauchy stress tensor and $\boldsymbol{g}$ is the body force per unit volume of fluid. In polymeric flow regimes the convection term (i.e. $v.\nabla v$) in Equation (1.2) is usually small and can be neglected. This is a characteristic of very low Reynolds number (creeping or Stokes) flow of highly viscous fluids. In the majority of polymer flow systems the body force in comparison to stress is very small and can also be omitted from the equation of motion.

The Cauchy stress tensor is given as

$$\boldsymbol{\sigma} = -p\boldsymbol{\delta} + \boldsymbol{\tau} \tag{1.3}$$

where p is hydrostatic pressure, $\boldsymbol{\delta}$ is unit second-order tensor (Kronecker delta) and $\boldsymbol{\tau}$ is the extra stress tensor. The equation of motion is hence written as

$$\rho\frac{\partial \boldsymbol{v}}{\partial t} + \boldsymbol{v} \cdot \nabla\boldsymbol{v} = -\nabla p\boldsymbol{\delta} + \nabla \cdot \boldsymbol{\tau} + \rho\boldsymbol{g} \tag{1.4}$$

1.1.3 Thermal energy equation

This equation is the expression of the conservation of thermal energy (first law of thermodynamics) and is written as

$$\rho c\frac{\mathrm{D}T}{\mathrm{D}t} = k\nabla^2 T + \boldsymbol{\tau} : \nabla\boldsymbol{v} + \dot{S} \tag{1.5}$$

where c is specific heat, k is thermal conductivity, T is temperature and ∇^2 is the scalar Laplacian. The terms on the right-hand side of Equation (1.5) represent heat flux due to conduction, viscous heat dissipation and a heat source (e.g. heat generated by chemical reactions etc.), respectively. Thermal energy changes related to the variations of fluid density are neglected in this equation. In processes involving solidification or melting of polymers, the specific heat varies substantially with temperature and it should be retained inside the material derivative in Equation (1.5). Thermal conductivity of polymers is also likely to be temperature dependent and anisotropic and, ideally, should be treated as a variable in the derivation of the energy equation. In practice, however, the lack of experimental data usually prevents the use of a variable k in polymer flow models.

1.1.4 Constitutive equations

A constitutive equation is a relation between the extra stress ($\boldsymbol{\tau}$) and the rate of deformation that a fluid experiences as it flows. Therefore, theoretically, the constitutive equation of a fluid characterises its macroscopic deformation behaviour under different flow conditions. It is reasonable to assume that the macroscopic behaviour of a fluid mainly depends on its microscopic structure. However, it is extremely difficult, if not impossible, to establish exact quantitative relationships between the microscopic structure of non-Newtonian fluids and their macroscopic properties. Therefore the derivation of universally applicable constitutive models for non-Newtonian fluids is, in general, not attempted. Instead, semi-empirical relationships which give a reasonable prediction for the behaviour of specified classes of non-Newtonian fluids under given flow conditions are used.

Depending on their constitutive behaviour, polymeric liquids are classified as:

- time-independent inelastic,
- time-dependent inelastic, or
- viscoelastic

non-Newtonian fluids.

1.2   CLASSIFICATION OF INELASTIC TIME-INDEPENDENT FLUIDS

In an inelastic, time-independent (Stokesian) fluid the extra stress is considered to be a function of the instantaneous rate of deformation (rate of strain). Therefore in this case the fluid does not retain any memory of the history of the deformation which it has experienced at previous stages of the flow.

1.2.1   Newtonian fluids

In the simplest case of Newtonian fluids (linear Stokesian fluids) the extra stress tensor is expressed, using a constant fluid viscosity μ, as

$$\boldsymbol{\tau} = 2\mu\boldsymbol{D} \tag{1.6}$$

where $\boldsymbol{D}$ is the rate of deformation (rate of strain) tensor representing the symmetric part of the velocity gradient tensor. Components of the rate of deformation tensor are hence given in terms of the velocity gradients as

$$D_{ij} = \tfrac{1}{2}(v_{i,j} + v_{j,i}) \tag{1.7}$$

Equations (1.6) and (1.7) are used to formulate explicit relationships between the extra stress components and the velocity gradients. Using these relationships the extra stress, $\boldsymbol{\tau}$, can be eliminated from the governing equations. This is the basis for the derivation of the well-known Navier–Stokes equations which represent the Newtonian flow (Aris, 1989).

1.2.2 Generalized Newtonian fluids

Common experimental evidence shows that the viscosity of polymers varies as they flow. Under certain conditions however, elastic effects in a polymeric flow can be neglected. In these situations the extra stress is again expressed, explicitly, in terms of the rate of deformation as

$$\boldsymbol{\tau} = 2\eta \boldsymbol{D} \tag{1.8}$$

where η is the apparent viscosity, which is a function of the magnitude of the rate of deformation tensor and temperature. Equation (1.8) is said to provide a 'generalized Newtonian' description of the fluid behaviour. Analogous to constant viscosity Newtonian flow, this equation is used to derive the 'generalized Navier–Stokes' equations via the substitution of the extra stress in the equation of motion in terms of viscosity and velocity gradients. Hence, the only requirement for the solution of these equations is the determination of the apparent fluid viscosity.

Theoretically the apparent viscosity of generalized Newtonian fluids can be found using a simple shear flow (i.e. steady state, one-dimensional, constant shear stress). The rate of deformation tensor in a simple shear flow is given as

$$\boldsymbol{D} = \dot{\gamma} \begin{pmatrix} 0 & \frac{1}{2} & 0 \\ \frac{1}{2} & 0 & 0 \\ 0 & 0 & 0 \end{pmatrix} \tag{1.9}$$

where $\dot{\gamma}$ is a scalar called the shear rate. Consequently in this case an experimental flow curve which relates the shear stress to the shear rate (called a rheogram) can be used to calculate the fluid viscosity as

$$\eta = \frac{\tau_{12}}{\dot{\gamma}} \tag{1.10}$$

where τ_{12} is the shear stress. In practice, it is very difficult to establish a simple shear flow and instead 'viscometric' regimes are used to determine parameters such as viscosity. In a viscometric flow, a fluid element deformation observed in a frame of reference which translates and rotates with that element, will be identical to a simple shear system.

In addition to the apparent viscosity two other material parameters can be obtained using simple shear flow viscometry. These are primary and secondary normal stress coefficients expressed, respectively, as

$$\Psi_{12} = \frac{\tau_{11} - \tau_{22}}{\dot{\gamma}^2} \tag{1.11}$$

and

$$\Psi_{23} = \frac{\tau_{22} - \tau_{33}}{\dot{\gamma}^2} \qquad\qquad (1.12)$$

Material parameters defined by Equations (1.11) and (1.12) arise from aniso-tropy (i.e. direction dependency) of the microstructure of long-chain polymers subjected to high shear deformations. Generalized Newtonian constitutive equations cannot predict any normal stress acting along the direction perpendicular to the shearing surface in a viscometric flow. Thus the primary and secondary normal stress coefficients are only used in conjunction with visco-elastic constitutive models.

Numerous examples of polymer flow models based on generalized Newtonian behaviour are found in non-Newtonian fluid mechanics literature. Using experimental evidence the time-independent generalized Newtonian fluids are divided into three groups. These are Bingham plastics, pseudoplastic fluids and dilatant fluids.

Bingham plastics

Bingham plastics are fluids which remain rigid under the application of shear stresses less than a yield stress, τ_y, but flow like a simple Newtonian fluid once the applied shear exceeds this value. Different constitutive models representing this type of fluids were developed by Herschel and Bulkley (1926), Oldroyd (1947) and Casson (1959).

Pseudoplastic fluids

Pseudoplastic fluids have no yield stress threshold and in these fluids the ratio of shear stress to the rate of shear generally falls continuously and rapidly with increase in the shear rate. Very low and very high shear regions are the exceptions, where the flow curve is almost horizontal (Figure 1.1).

A common choice of functional relationship between shear viscosity and shear rate, that usually gives a good prediction for the shear thinning region in pseudoplastic fluids, is the power law model proposed by de Waele (1923) and Ostwald (1925). This model is written as the following equation

$$\eta = \eta_0(\dot{\gamma})^{n-1} \qquad\qquad (1.13)$$

where the parameters, η_0 and n, are called the consistency coefficient and the power law index, respectively. It is clear that a fluid with power law index of unity will be a purely Newtonian fluid. It is also commonly accepted that the non-Newtonian behaviour of fluids become more pronounced as their corresponding

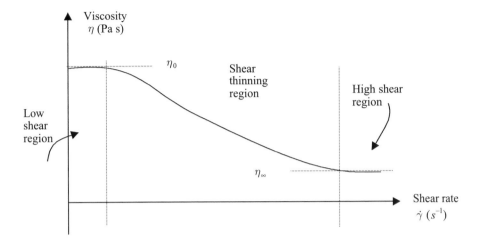

Figure 1.1   Shear thinning behaviour of pseudoplastic fluids

power law index shows greater departure from unity. The consistency coefficient is fluid viscosity at zero shear and it has a higher value for more viscous fluids. A modified version of the power law model which can represent very low shear regions has also been proposed (Middleman, 1977). In some cases it may be more realistic to apply a segmented form of this model in which different values of the parameters over different ranges of the shear rate are used. The following temperature-dependent form of the power law equation, based on the Arrhenius formula for thermal effects, is the most frequently used version of this relation in polymer flow models (Pittman and Nakazawa, 1984)

$$\eta = \eta_0(\dot{\gamma})^{n-1}e^{-b(T-T_{\text{ref}})} \tag{1.14}$$

where b is called the temperature dependency coefficient and T_{ref} is a reference temperature. In addition to the power law model a plethora of other relationships representing the constitutive behaviour of pseudoplastic fluids can also be found in the literature. For example, the following equation proposed by Carreau (1968) has gained widespread application in polymer processing analysis

$$\frac{\eta - \eta_\infty}{\eta_0 - \eta_\infty} = [1 + (\lambda|\dot{\gamma}|)^2]^{(n-1)/2} \tag{1.15}$$

where η_0 and η_∞ are zero and infinite shear rate 'constant viscosities', respectively, λ is a material time constant and n is the power law index.

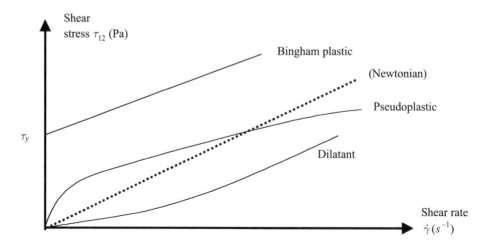

Figure 1.2 Comparison of the rheological behaviour of Newtonian and typical generalized Newtonian fluids

Dilatant fluids

Dilatant fluids (also known as shear thickening fluids) show an increase in viscosity with an increase in shear rate. Such an increase in viscosity may, or may not, be accompanied by a measurable change in the volume of the fluid (Metzener and Whitlock, 1958). Power law-type rheological equations with $n > 1$ are usually used to model this type of fluids.

Typical rheograms representing the behaviour of various types of generalized Newtonian fluids are shown in Figure 1.2.

1.3 INELASTIC TIME-DEPENDENT FLUIDS

Under the application of a steady rate of shear, the viscosity of some types of non-Newtonian fluids changes with time. Time-dependent fluids, which show an increase in viscosity as time passes, are called 'rheopectic'. Fluids showing the opposite effect of decreasing viscosity are called 'thixotropic'. Rheopexy and thixotropy are complex phenomena resulting from transient changes of the molecular structure of time-dependent fluids under an applied shear stress. In general, it is extremely difficult to introduce molecular effects of this kind into the constitutive equations of non-Newtonian fluids. Thus the proposed constitutive models for these fluids are based on many simplifying assumptions (Slibar and Paslay, 1959). In cases where the elastic effects shown by a time-dependent fluid are negligible then, basically, a mathematical model similar to the generalized Newtonian fluids can be used to represent their flow. The constitutive equation in such a model must, however, reflect the time dependency of the fluid viscosity. The construction of flow models for time-dependent fluids often requires the use of kinetical relations. These relations represent molecular phenomenon such as polymer degradation as a function of time (Kemblowski and Petera, 1981).

1.4   VISCOELASTIC FLUIDS

Apart from the prediction of a variable viscosity, generalized Newtonian constitutive models cannot explain other phenomena such as recoil, stress relaxation, stress overshoot and extrudate swell which are commonly observed in polymer processing flows. These effects have a significant impact on the product quality in polymer processing and they should not be ignored. Theoretically, all of these phenomena can be considered as the result of the material having a combination of the properties of elastic solids and viscous fluids. Therefore mathematical modelling of polymer processing flows should, ideally, be based on the use of viscoelastic constitutive equations. Formulation of the constitutive equations for viscoelastic fluids has been the subject of a considerable amount of research over many decades. Details of the derivation of the viscoelastic constitutive equations and their classification are covered in many textbooks and review papers (see Tanner, 1985; Bird *et al.*, 1977; Mitsoulis, 1990). Despite these efforts and the proliferation of proposed viscoelastic constitutive equations in recent years, the problem of selecting one which can yield verifiable results for a fluid under all types of flow conditions is still unresolved (Pearson, 1994). In practice, therefore, the remaining option is to choose a constitutive viscoelastic model that can predict the most dominant features of the fluid behaviour for a given flow situation. It should also be mentioned here that the use of a computationally costly and complex viscoelastic model in situations that are different from those assumed in the formulation of that model will in general yield unreliable predictions and should be avoided.

1.4.1   Model (material) parameters used in viscoelastic constitutive equations

Material parameters, such as relaxation time, elongational viscosity and normal stress coefficients, are essentially used as convenient means of introducing various aspects of viscoelastic fluid behaviour into a constitutive equation. The exact definition of these parameters under general conditions is difficult and hence they are regarded as empirical parameters in viscoelastic models. These parameters can, however, be regarded as compatible with well-defined physical functions under simple flow conditions. Thus, analogous to the fluid viscosity in Newtonian flows, the material parameters in a viscoelastic model are found via rheometric experiments conducted using simple flow regimes.

Stress relaxation time, obtained from rheograms based on viscometric flows, is used to define a dimensionless parameter called the 'Deborah number', which quantifies the elastic character of a fluid

$$De = \frac{\lambda}{\theta} = \lambda \dot{\gamma} \tag{1.16}$$

where λ is the relaxation time (characteristic of polymer chains relaxation) and θ is an appropriate duration of the deformation time. The magnitude of the Deborah number is used as a measure for deciding whether viscoelastic effects in a certain flow problem are significant or not. An alternative definition based on the dimensionless 'Weissenberg number' is also used to provide a quantitative measure for viscoelasticity of non-Newtonian fluids (Middleman, 1977). The Weissenberg number is defined as

$$W_s = \lambda \frac{V}{H} \tag{1.17}$$

where V and H are the characteristic process velocity and process length, respectively.

The rate of deformation tensor in a pure elongational flow has the following form

$$\mathbf{D} = \dot{\varepsilon} \begin{pmatrix} 1 & 0 & 0 \\ 0 & -\frac{1}{2} & 0 \\ 0 & 0 & -\frac{1}{2} \end{pmatrix} \tag{1.18}$$

where $\dot{\varepsilon}$ is a positive scalar called the principal extension rate. Based on a pure elongational flow the elongational viscosity of a fluid is defined as

$$\eta_e = \frac{\tau_{11} - \tau_{22}}{\dot{\varepsilon}} \tag{1.19}$$

where τ_{11} and τ_{22} are the normal stress components. In practice, it is very difficult to set up a controlled, pure elongational flow and the measurement of the elongational viscosity of fluids is not a trivial matter. For a Newtonian fluid the elongational viscosity is three times the value of μ (Stevenson, 1972). In viscoelastic fluids the ratio of elongational viscosity to shear viscosity can be much higher than three.

The elongation viscosity defined by Equation (1.19) represents a uni-axial extension. Elongational flows based on biaxial extensions can also be considered. In an equi-biaxial extension the rate of deformation tensor is defined as

$$D = \dot{\varepsilon}_B \begin{pmatrix} 1 & 0 & 0 \\ 0 & 1 & 0 \\ 0 & 0 & -2 \end{pmatrix} \tag{1.20}$$

where $\dot{\varepsilon}_B$ is a positive scalar called the biaxial elongation rate. The biaxial viscosity is defined as

$$\eta_B = \frac{\tau_{11} - \tau_{33}}{\dot{\varepsilon}_B} = \frac{\tau_{22} - \tau_{33}}{\dot{\varepsilon}_B} \tag{1.21}$$

1.4.2 Differential constitutive equations for viscoelastic fluids

Depending on the method of analysis, constitutive models of viscoelastic fluids can be formulated as differential or integral equations.

In the differential models stress components, and their material derivatives, are related to the rate of strain components and their material derivatives.

The Oldroyd-type differential constitutive equations for incompressible viscoelastic fluids can in general can be written as (Oldroyd, 1950)

$$\lambda \frac{\Delta_{a,b,c}\tau}{\Delta t} + \tau = 2\eta \left(D + \Lambda \frac{\Delta_{a,b,c}D}{\Delta t} \right) \tag{1.22}$$

where λ, η and Λ are material parameters and the time derivative $(\Delta_{a,b,c}Y)/\Delta t$ of a tensor Y is defined as

$$\lambda \frac{\Delta_{a,b,c}Y}{\Delta t} = \frac{\partial Y}{\partial t} + v.\nabla Y + (\omega.Y - Y.\omega) + a(D.Y + Y.D)$$

$$+ bI \operatorname{trace}(D.Y) + cD \operatorname{trace} Y \tag{1.23}$$

where I is the unit tensor, $D = \frac{1}{2}[\nabla v + (\nabla v)^T]$ and $\omega = \frac{1}{2}[\nabla v - (\nabla v)^T]$ are the symmetric and antisymmetric parts of the velocity gradient tensor. Equation (1.23) gives the most general definition of a time derivative of any second-order tensor and it contains the local, convective, rotational and strain related changes of the tensor with respect to the time variable. In practice special cases are considered. The Jaumann or co-rotational time derivative is defined as a case where the parameters a, b and c are zero. In the upper-convected (or co-deformational) Oldroyd derivative, $a = -1$ and b and c are zero. The lower-convected Oldroyd derivative is defined using $a = +1$ and b and c as zeros.

The Maxwell class of viscoelastic constitutive equations are described by a simpler form of Equation (1.22) in which $\Lambda = 0$. For example, the upper-convected Maxwell model (UCM) is expressed as

$$\boldsymbol{\tau} + \lambda \frac{\Delta_{-100}\boldsymbol{\tau}}{\Delta t} = 2\eta \boldsymbol{D} \tag{1.24}$$

Other combinations of upper- and lower-convected time derivatives of the stress tensor are also used to construct constitutive equations for viscoelastic fluids. For example, Johnson and Segalman (1977) have proposed the following equation

$$\boldsymbol{\tau} + \lambda \left[\left(1 - \frac{\varsigma}{2}\right) \frac{\Delta_{-100}\boldsymbol{\tau}}{\Delta t} + \frac{\varsigma}{2} \frac{\Delta_{100}\boldsymbol{\tau}}{\Delta t} \right] = 2\eta \boldsymbol{D} \tag{1.25}$$

where ς is a parameter between 0 and 2.

A frequently used example of Oldroyd-type constitutive equations is the Oldroyd-B model. The Oldroyd-B model can be thought of as a description of the constitutive behaviour of a fluid made by the dissolution of a (UCM) fluid in a Newtonian 'solvent'. Here, the parameter Λ, called the 'retardation time' is defined as $\Lambda = \lambda \left(\eta_s/(\eta + \eta_s)\right)$, where η_s is the viscosity of the solvent. Hence the extra stress tensor in the Oldroyd-B model is made up of Maxwell and solvent contributions. The Oldroyd-B constitutive equation is written as

$$\boldsymbol{\tau} + \lambda \frac{\Delta_{-100}\boldsymbol{\tau}}{\Delta t} = 2(\eta + \eta_s)\left(\boldsymbol{D} + \lambda \frac{\eta_s}{\eta + \eta_s}\frac{\Delta_{-100}\boldsymbol{D}}{\Delta t}\right) \tag{1.26}$$

In general, Maxwell or Oldroyd models do not give realistic predictions of the flow and deformation behaviour of polymeric fluids. In particular, in cases where the flow regime is characterized by elongational deformations these models are found to give very poor predictions. There have been many attempts to derive constitutive models that incorporate both shear and elongational behaviour of viscoelastic fluids. Phan-Thien and Tanner (1977) formulated a viscoelastic model based on the network theory for macromolecules. This model has been shown to give relatively good results for elongational flows (Tanner, 1985). The Phan-Thien/Tanner equation is expressed as

$$\exp\left(\frac{\varepsilon\lambda}{\eta}\mathrm{trace}\ \boldsymbol{\tau}\right)\boldsymbol{\tau} + \lambda \left[\left(1 - \frac{\varsigma}{2}\right)\frac{\Delta_{-100}\boldsymbol{\tau}}{\Delta t} + \frac{\varsigma}{2}\frac{\Delta_{100}\boldsymbol{\tau}}{\Delta t}\right] = 2\eta \boldsymbol{D} \tag{1.27}$$

where ε is defined as a characteristic elongational parameter. In Equation (1.27) the parameters ε and ς ($0 \leq \varsigma \leq 2$) are representative of the elongational and shear behaviour of the fluid, respectively. As it can be seen the insertion of $\varepsilon = 0$ reduces the Phan-Thien/Tanner equation to the Johnson/Segalman model.

All of the described differential viscoelastic constitutive equations are implicit relations between the extra stress and the rate of deformation tensors. Therefore, unlike the generalized Newtonian flows, these equations cannot be used to eliminate the extra stress in the equation of motion and should be solved 'simultaneously' with the governing flow equations.

1.4.3   Single-integral constitutive equations for viscoelastic fluids

In integral models, stress components are obtained by integrating appropriate functions, representing the amount of deformation, over the strain history of the fluid. The simplest integral constitutive model for rubber-like fluids was proposed by Lodge (1964). Other equations, belonging to this category of constitutive models, have been developed by Johnson and Segalman (1977) and Doi and Edwards (1978, 1979). However, the most frequently used single-integral constitutive model for viscoelastic fluids is the KBKZ equation, independently proposed by Kaye (1962) and Bernstein, Kearsley and Zapas (1963). The generic form of the integral constitutive equations for isotropic fluids is written as

$$\boldsymbol{\tau}(t) = \int_{-\infty}^{t} M(t - t')\{\Phi_1(I_1, I_2)[\boldsymbol{F}_t(t - t') - \boldsymbol{I}] + \Phi_2(I_1, I_2)[\boldsymbol{C}_t(t - t') - \boldsymbol{I}]\} \mathrm{d}t' \quad (1.28)$$

Here $\boldsymbol{\tau}(t)$ is the stress at a fluid particle given by an integral of deformation history along the fluid particle trajectory between a deformed configuration at time t' and the current reference time t.

In Equation (1.28) function $M(t - t')$ is the time-dependent memory function of linear viscoelasticity, non-dimensional scalars Φ_1 and Φ_2 and are the functions of the first invariant of $\boldsymbol{C}_t(t - t')$ and $\boldsymbol{F}_t(t - t')$, which are, respectively, the right Cauchy–Green tensor and its inverse (called the Finger strain tensor) (Mitsoulis, 1990). The memory function is usually expressed as

$$M(t - t') = \sum_{k=1}^{n} \frac{\eta_k}{\lambda_k^2} \exp\left[\frac{-(t - t')}{\lambda_k}\right] \quad (1.29)$$

where $\eta_k (k = 1, n)$ are the viscosity coefficients and $\lambda_k (k = 1, n)$ are the relaxation times. The general Equation (1.27) can be used to derive various single-integral viscoelastic constitutive models for incompressible fluids. For example, by setting $\Phi_1 = 1$ and $\Phi_2 = 0$ the model developed by Lodge (1964) is derived. This model can be shown to be equivalent to the upper-convected Maxwell equation described in the previous section. To obtain the KBKZ model the scalar functions in the strain-dependent kernel of the integral in Equation (1.27) are chosen as

$$\Phi_1 = \frac{\partial W(I_1, I_2)}{\partial I_1} \quad \text{and} \quad \Phi_2 = -\frac{\partial W(I_1, I_2)}{\partial I_2} \quad (1.30)$$

where I_1 and $\mathrm{tr}(\boldsymbol{F})$ and $I_2 = \mathrm{tr}(\boldsymbol{C})$ and $W(I_1, I_2)$ is a potential for functions Φ_1 and Φ_2. As it can be seen the Lodge model is a special case of the KBKZ model in which $W = I_1$. A simplified form of the KBKZ model proposed by Wagner (1979) has found widespread application in the modelling of viscoelastic flows (see Olley and Coates, 1997).

Integral models have the apparent advantage of giving the extra stress tensor explicitly and therefore they can be used to find the stress in a separate step to other field unknowns. However, the integral models are mathematically difficult to handle and, in general, should be solved in Lagrangian frameworks. The high computational costs of adaptive meshing required in Lagrangian systems and problems arising from the calculation of functions which are dependent on strain history are regarded as the set backs for these models.

Some of the integral or differential constitutive equations presented in this and the previous section have an exact equivalent in the other group. There are, however, equations in both groups that have no equivalent in the other category.

1.4.4   Viscometric approach – the (CEF) model

The practical and computational complications encountered in obtaining solutions for the described differential or integral viscoelastic equations sometimes justifies using a heuristic approach based on an equation proposed by Criminale, Ericksen and Filbey (1958) to model polymer flows. Similar to the generalized Newtonian approach, under steady-state viscometric flow conditions components of the extra stress in the (CEF) model are given as explicit relationships in terms of the components of the rate of deformation tensor. However, in the (CEF) model stress components are 'corrected' to take into account the influence of normal stresses in non-Newtonian flow behaviour. For example, in a two-dimensional planar coordinate system the components of extra stress in the (CEF) model are written as

$$
\begin{cases}
\tau_{xx} = \eta D_{xx} + (\Psi_{12} + \Psi_{23})D_{xy}^2 \\
\tau_{yy} = \eta D_{yy} + \Psi_{23}D_{xy} \\
\tau_{xy} = \eta D_{xy}
\end{cases}
\tag{1.31}
$$

where D_{xx} etc. are the components of the rate of deformation (strain) tensor and Ψ_{12} and Ψ_{23} are the primary and secondary normal stress coefficients, respectively. An analogous set of relationships which reflect elongational behaviour of polymeric liquids has also been proposed (Mitsoulis, 1990) as

$$
\begin{cases}
\tau_{xx} = \eta_e D_{xx} + (\Psi_{12} + \Psi_{23})D_{xy}^2 \\
\tau_{yy} = \eta_e D_{yy} + \Psi_{23}D_{xy} \\
\tau_{xy} = \eta_s D_{xy}
\end{cases}
\tag{1.32}
$$

where η_e and η_s are elongational and shear viscosity, respectively. Due to the low computational costs of using the (CEF) model, this approach has been advocated as an attractive alternative to more complex viscoelastic equations in the modelling of polymer flow systems that do not significantly deviate from viscometric conditions (Mitsoulis, 1986).

REFERENCES

Aris, R., 1989. *Vectors, Tensors and the Basic Equations of Fluid Mechanics*, Dover Publications, New York.

Bernstein, B., Kearsley, E.A. and Zapas, L., 1963. A study of stress relaxation with finite strain. *Trans. Soc. Rheol.* **7**, 391–410.

Bird, R.B., Armstrong, R.C. and Hassager, O., 1977. *Dynamics of Polymeric Fluids, Vol. 1: Fluid Mechanics*, Wiley, New York.

Bird, R.B., Stewart, W.E. and Lightfoot, E.M., 1960. *Transport Phenomena*, Wiley, New York.

Casson, N., 1959. In: Mill. C.C. (ed.), *Rheology of Disperse Systems*, Pergamon Press, London.

Carreau, P.J., 1968. PhD thesis, Department of Chemical Engineering, University of Wisconsin, Wisconsin.

Criminale, W.O. Jr, Ericksen, J.L. and Filby, G.L. Jr., 1958. Steady shear flow of non-Newtonian fluids. *Arch. Rat. Mech. Anal.* **1**, 410–417.

de Waele, A., 1923. See Bird, R.B., Armstrong, R.C. and Hassager, O. 1977. *Dynamics of Polymeric Fluids, Vol. 1: Fluid Mechanics*, Wiley, New York.

Doi, M. and Edwards, S.F., 1978. Dynamics of concentrated polymer systems: 1. Brownian motion in equilibrium state, 2. Molecular motion under flow, 3. Constitutive equation and 4. Rheological properties. *J. Chem. Soc., Faraday Trans. 2* **74**, 1789, 1802, 1818–1832.

Doi, M. and Edwards, S.F., 1979. Dynamics of concentrated polymer systems: 1. Brownian motion in equilibrium state, 2. Molecular motion under flow, 3. Constitutive equation and 4. Rheological properties. *J. Chem. Soc., Faraday Trans. 2* **75**, 38–54.

Herschel, W.H. and Bulkley, R., 1927. See Rudraiah, N. and Kaloni, P.N. 1990. Flow of non-Newtonian fluids. In: *Encyclopaedia of Fluid Mechanics*, Vol. 9, Chapter 1, Gulf Publishers, Houston.

Johnson, M.W. and Segalman, D., 1977. A model for viscoelastic fluid behaviour which allows non-affine deformation. *J. Non-Newtonian Fluid Mech.* **2**, 255–270.

Kaye, A., 1962. *Non-Newtonian Flow in Incompressible Fluids*, CoA Note No. 134, College of Aeronautics, Cranfield.

Kemblowski, Z. and Petera, J., 1981. Memory effects during the flow of thixotropic fluids in pipes. *Rheol. Acta* **20**, 311–323.

Lodge, A.S., 1964. *Elastic Liquids*, Academic Press, London.

Metzener, A.B. and Whitlock, M., 1958. Flow behaviour of concentrated dilatant suspensions. *Trans. Soc. Rheol.* **2**, 239–254.

Middleman, S., 1977. *Fundamentals of Polymer Processing*, McGraw-Hill, New York.

Mitsoulis, E., 1986. The numerical simulation of Boger fluids: a viscometric approximation approach. *Polym. Eng. Sci.* **26**, 1552–1562.

Mitsoulis, E., 1990. Numerical Simulation of Viscoelastic Fluids. In: *Encyclopaedia of Fluid Mechanics*, Vol. 9, Chapter 21, Gulf Publishers, Houston.

Oldroyd, J.G., 1947. A rational formulation of the equations of plastic flow for a Bingham solid. *Proc. Camb. Philos. Soc.* **43**, 100–105.

Oldroyd, J.G., 1950. On the formulation of rheological equations of state. *Proc. Roy. Soc.* **A200**, 523–541.

Olley, P. and Coates, P. D., 1997. An approximation to the KBKZ constitutive equation. *J. Non-Newtonian Fluid Mech.* **69**, 239–254.

Ostwald, W., 1925. See Bird, R. B., Armstrong, R. C. and Hassager, O. 1977. *Dynamics of Polymeric Fluids, Vol. 1: Fluid Mechanics*, Wiley, New York.

Pearson, J. R. A., 1994. Report on University of Wales Institute of Non-Newtonian Fluid Mechanics Mini Symposium on Continuum and Microstructural Modelling in Computational Rheology. *J. Non-Newtonian Fluid Mech.* **55**, 203–205.

Phan-Thien, N. and Tanner, R. I., 1977. A new constitutive equation derived from network theory. *J. Non-Newtonian Fluid Mech.* **2**, 353–365.

Pittman, J. F. T. and Nakazawa, S., 1984. Finite element analysis of polymer processing operations. In: Pittman, J. F. T., Zienkiewicz, O. C., Wood, R. D. and Alexander, J. M. (eds), *Numerical Analysis of Forming Processes*, Wiley, Chichester.

Slibar, A. and Paslay, P. R., 1959. Retarded flow of Bingham materials. *Trans. ASME* **26**, 107–113.

Stevenson, J. F., 1972. Elongational flow of polymer melts. *AIChE. J.* **18**, 540–547.

Tadmor, Z. and Gogos, C. G., 1979. *Principles of Polymer Processing*, Wiley, New York.

Tanner, R. I., 1985. *Engineering Rheology*, Clarendon Press, Oxford.

Wagner, M. H., 1979. Towards a network theory for polymer melts. *Rheol. Acta.* **18**, 33–50.

2

Weighted Residual Finite Element Methods – an Outline

As described in Chapter 1, mathematical models that represent polymer flow systems are, in general, based on non-linear partial differential equations and cannot be solved by analytical techniques. Therefore, in general, these equations are solved using numerical methods. Numerical solutions of the differential equations arising in engineering problems are usually based on finite difference, finite element, boundary element or finite volume schemes. Other numerical techniques such as the spectral expansions or newly emerged mesh independent methods may also be used to solve governing equations of specific types of engineering problems. Numerous examples of the successful application of these methods in the computer modelling of realistic field problems can be found in the literature. All of these methods have strengths and weaknesses and a number of factors should be considered before deciding in favour of the application of a particular method to the modelling of a process. The most important factors in this respect, are: type of the governing equations of the process, geometry of the process domain, nature of the boundary conditions, required accuracy of the calculations and computational cost.

In general, the finite element method has a greater geometrical flexibility than other currently available numerical techniques. It can also cope very effectively with various types of boundary conditions. The most significant setback for this method is the high computational cost of three-dimensional finite element models. In practice, rational approximations are often used to obtain useful simulations for realistic problems without full three-dimensional analysis. In the finite element modelling of polymeric flows the following approaches can be adopted to achieve computing economy:

- Two-dimensional models can be used to provide effective approximations in the modelling of polymer processes if the flow field variations in the remaining (third) direction are small. In particular, in axisymmetric domains it may be possible to ignore the circumferential variations of the field unknowns and analytically integrate the flow equations in that direction to reduce the numerical model to a two-dimensional form.

- Process characteristics may justify the use of simplifying assumptions such as the 'lubrication approximation' which may be applied to represent creeping flow in narrow gaps.

- Components of the governing equations of the process can be decoupled to develop a solution scheme for a three-dimensional problem by combining one- and two-dimensional analyses.

Examples of polymeric flow models where the above simplifications have been successfully used are presented in Chapter 5.

Finite element modelling of engineering processes can be based on different methodologies. For example, the preferred method in structural analyses is the 'displacement method' which is based on the minimization of a variational statement that represents the state of equilibrium in a structure (Zienkiewicz and Taylor, 1994). Engineering fluid flow processes, on the other hand, cannot be usually expressed in terms of variational principles. Therefore, the mathematical modelling of fluid dynamical problems is mainly dependent on the solution of partial differential equations derived from the laws of conservation of mass, momentum and energy and constitutive equations. Weighted residual methods, such as the Galerkin, least square and collocation techniques provide theoretical basis for the numerical solution of partial differential equations. However, the direct application of these techniques to engineering problems is usually not practical and they need to be combined with finite element approximation procedures to develop robust practical schemes. Hence the commonly adopted approach in computer modelling of flow processes in polymer engineering operations is the application of weighted residual finite element methods.

The main concepts of the finite element approximation and the general outline of the weighted residual methods are briefly explained in this chapter. These concepts provide the necessary background for the development of the working equations of the numerical schemes used in the simulation of polymer processing operations. In-depth analyses of the mathematical theorems underpinning finite element approximations and weighted residual methods are outside the scope of this book. Therefore, in this chapter, mainly descriptive outlines of these topics are given. Detailed explanations of the theoretical aspects of the solution of partial differential equations by the weighted residual finite element methods can be found in many textbooks dedicated to these subjects. For example, see Mitchell and Wait (1977), Johnson (1987), Brenner and Scott (1994) and, specifically, for the solution of incompressible Navier–Stokes equations see Girault and Raviart (1986) and Pironneau (1989).

Mathematical derivations presented in the following sections are, occasionally, given in the context of one- or two-dimensional Cartesian coordinate systems. These derivations can, however, be readily generalized and the adopted style is to make the explanations as simple as possible.

2.1 FINITE ELEMENT APPROXIMATION

The first step in the formulation of a finite element approximation for a field problem is to divide the problem domain into a number of smaller sub-regions without leaving any gaps or overlapping between them. This process is called 'Domain Discretization'. An individual sub-region in a discretized domain is called a 'finite element' and collectively, the finite elements provide a 'finite element mesh' for the discretized domain. In general, the elements in a finite element mesh may have different sizes but all of them usually have a common basic shape (e.g. they are all triangular or quadrilateral) and an equal number of nodes. The nodes are the sampling points in an element where the numerical values of the unknowns are to be calculated. All types of finite elements should have some nodes located on their boundary lines. Some of the commonly used finite elements also have interior nodes. Boundary nodes of the individual finite elements appear as the junction points between the elements in a finite element mesh. In the finite element modelling of flow processes the elements in the computational mesh are geometrical sub-regions of the flow domain and they do not represent parts of the body of the fluid.

In most engineering problems the boundary of the problem domain includes curved sections. The discretization of domains with curved boundaries using meshes that consist of elements with straight sides inevitably involves some error. This type of discretization error can obviously be reduced by mesh refinements. However, in general, it cannot be entirely eliminated unless finite elements which themselves have curved sides are used.

The discretization of a problem domain into a finite element mesh consisting of randomly sized triangular elements is shown in Figure 2.1. In the coarse mesh shown there are relatively large gaps between the actual domain boundary and the boundary of the mesh and hence the overall discretization error is expected to be large.

The main consequence of the discretization of a problem domain into finite elements is that within each element, unknown functions can be approximated using interpolation procedures.

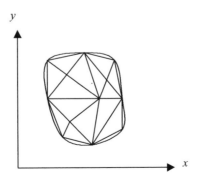

Figure 2.1 Problem domain discretization

2.1.1 Interpolation models

Let Ω_e be a well-defined finite element, i.e. its shape, size and the number and locations of its nodes are known. We seek to define the variations of a real valued continuous function, such as f, over this element in terms of appropriate geometrical functions. If it can be assumed that the values of f on the nodes of Ω_e are known, then in any other point within this element we can find an approximate value for f using an interpolation method. For example, consider a one-dimensional two-node (linear) element of length l with its nodes located at points $A(x_A = 0)$ and $B(x_B = l)$ as is shown in Figure 2.2.

Figure 2.2   A one-dimensional linear element

Using a simple interpolation procedure variations of a continuous function such as f along the element can be shown, approximately, as

$$\tilde{f}_x = \frac{l-x}{l}f_A + \frac{x}{l}f_B \tag{2.1}$$

Equation (2.1) provides an approximate interpolated value for f at position x in terms of its nodal values and two geometrical functions. The geometrical functions in Equation (2.1) are called the 'shape' functions. A simple inspection shows that: (a) each function is equal to 1 at its associated node and is 0 at the other node, and (b) the sum of the shape functions is equal to 1. These functions, shown in Figure 2.3, are written according to their associated nodes as N_A and N_B.

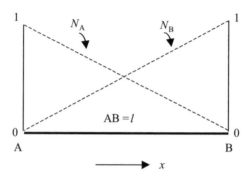

Figure 2.3   Linear Lagrange interpolation functions

$$\left(N_{\mathrm{A}} = \frac{l-x}{l} \text{ and } N_{\mathrm{B}} = \frac{x}{l} \right)$$

Analogous interpolation procedures involving higher numbers of sampling points than the two ends used in the above example provide higher-order approximations for unknown functions over one-dimensional elements. The method can also be extended to two- and three-dimensional elements. In general, an interpolated function over a multi-dimensional element Ω_e is expressed as

$$f(\bar{x}) \approx \tilde{f}(\bar{x}) = \sum_{i=1}^{p} N_i(\bar{x}) f_i \tag{2.2}$$

where $(\bar{x})$ represents the coordinates of the point in Ω_e on which we wish to find an approximate (interpolated) value for the function f, i is the node index, p is the total number of nodes in Ω_e and $N_i(\bar{x})$ is the shape function associated with node i. Similar to the one-dimensional example the shape functions in multi-dimensional elements should also satisfy the following conditions

$$\begin{cases} N_i(\bar{x}_j) = 1, & i = j \\ N_i(\bar{x}_j) = 0, & i \neq j \end{cases} \text{ and } \sum_{i=1}^{p} N_i(\bar{x}_j) = 1 \tag{2.3}$$

where f_i, $i = 1, \ldots p$ are the nodal values of the function f (called the nodal degrees of freedom). Nodal degrees of freedom appearing in elemental interpolations (i.e. f_i, $i = 1, \ldots p$) are the field unknowns that will be found during the finite element solution procedure. The general form of the shape functions associated with a finite element depends on its shape and the number of its nodes. In most types of commonly used finite elements these functions are low degree polynomials. In general, if the degrees of freedom in a finite element are all given as nodal values of unknown functions (i.e. function derivatives are excluded) then the element is said to belong to the Lagrange family of elements. However, some authors use the term 'Lagrange element' exclusively for those elements whose associated shape functions are specifically based on Lagrange interpolation polynomials or their products (Gerald and Wheatley, 1984).

Hermite interpolation models involving the derivatives of field variables (Ciarlet, 1978; Lapidus and Pinder, 1982) can also be used to construct function approximations over finite elements. Consider a two-node one-dimensional element, as is shown in Figure 2.4, in which the degrees of freedom are nodal values and slopes of unknown functions. Therefore the expression defining the approximate value of a function f at a point in the interior of this element should include both its nodal values and slopes. Let this be written as

$$\tilde{f} = \sum_{I=1}^{2} N_{0I}(x)f_I + N_{1I}(x)\frac{\partial f_I}{\partial x} \tag{2.4}$$

Figure 2.4 A one-dimensional Hermite element

where $N_{0I}(x)$ and $N_{1I}(x)$ are polynomial expansions of equal order. As it can be seen in this case each node is associated with two shape functions. At the ends of the line element Equation (2.4) must give the function values and slopes shown in Figure 2.4, therefore:

- $N_{0I}(x)$ must be 1 at node number I and 0 at the other node, $N'_{0I}(x)$ must be 0 at both nodes, and

- $N_{1I}(x)$ must be 0 at both nodes and $N'_{1I}(x)$ must be 1 at node I and 0 at the other node.

A simple inspection shows that cubic functions (splines) shown graphically in Figure 2.5 satisfy the above conditions.

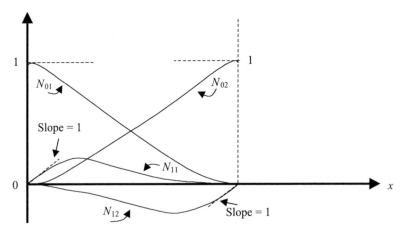

Figure 2.5 One-dimensional Hermite interpolation functions

Inherent in the development of approximations by the described interpolation models is to assign polynomial variations for function expansions over finite elements. Therefore the shape functions in a given finite element correspond to a particular approximating polynomial. However, finite element approximations may not represent complete polynomials of any given degree.

2.1.2 Shape functions of commonly used finite elements

Standard procedures for the derivation of the shape functions of common types of finite elements can be illustrated in the context of two-dimensional triangular and rectangular elements. Let us, first, consider a triangular element having three nodes located at its vertices as is shown in Figure 2.6.

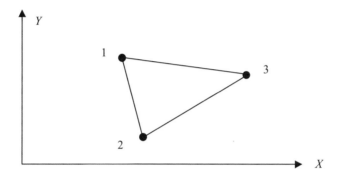

Figure 2.6   A linear triangular element

Variations of a continuous function over this element can be represented by a complete first-order (linear) polynomial as

$$f(x, y) \approx \tilde{f}(x, y) = a_1 + a_2 x + a_3 y \qquad (2.5)$$

By the insertion of the nodal coordinates into Equation (2.5) nodal values of f can be found. This is shown as

$$\begin{Bmatrix} f_1 \\ f_2 \\ f_3 \end{Bmatrix} = \begin{bmatrix} 1 & x_1 & y_1 \\ 1 & x_2 & y_2 \\ 1 & x_3 & y_3 \end{bmatrix} \begin{Bmatrix} a_1 \\ a_2 \\ a_3 \end{Bmatrix} \qquad (2.6)$$

where x_i, y_i, $i = 1,3$ are the nodal coordinates and f, $i = 1,3$ are the nodal degrees of freedom (i.e. function values). Using matrix notation Equation (2.6) is written as

$$f^e = Ca \qquad (2.7)$$

hence

$$a = C^{-1}f^e \qquad (2.8)$$

Equation (2.5) can be written as

$$\tilde{f} = Pa = PC^{-1}f^e \qquad (2.9)$$

where $P = [1 \ x \ y]$. Comparing Equations (2.2) and (2.9) we have

$$\tilde{f} = Nf^e \qquad (2.10)$$

where N is the set of shape functions written as

$$N = PC^{-1} \qquad (2.11)$$

In the outlined procedure the derivation of the shape functions of a three-noded (linear) triangular element requires the solution of a set of algebraic equations, generally shown as Equation (2.7).

Shape functions of the described triangular element are hence found on the basis of Equation (2.11) as

$$N_1 = \frac{(y_2 - y_3)x + (x_3 - x_2)y + (x_2y_3 - x_3y_2)}{x_1y_2 + x_2y_3 + x_3y_1 - x_1y_3 - x_2y_1 - x_3y_2} \qquad (2.12)$$

$$N_2 = \frac{(y_3 - y_1)x + (x_1 - x_3)y + (x_3y_1 - x_1y_3)}{x_1y_2 + x_2y_3 + x_3y_1 - x_1y_3 - x_2y_1 - x_3y_2} \qquad (2.13)$$

$$N_3 = \frac{(y_1 - y_2)x + (x_2 - x_1)y + (x_1y_2 - x_2y_1)}{x_1y_2 + x_2y_3 + x_3y_1 - x_1y_3 - x_2y_1 - x_3y_2} \qquad (2.14)$$

It can be readily shown that these geometric functions satisfy the conditions described by Equation (2.3).

Shape functions of a quadratic triangular element, with six associated nodes located at its vertices and mid-sides, can be derived by a similar procedure using a complete second order polynomial. Similarly it can be shown that a complete cubic polynomial corresponds to a triangular element with 10 nodes and so on. The arrangement shown in Figure 2.7 (called the Pascal triangle) represents the terms required to construct complete polynomials of any given degree, p, in two variables x and y.

The number of terms of a complete polynomial of any given degree will hence correspond to the number of nodes in a triangular element belonging to this family. An analogous tetrahedral family of finite elements that corresponds to complete polynomials in terms of three spatial variables can also be constructed for three-dimensional analysis.

$$1 \qquad\qquad\qquad\qquad p = 0$$

$$x \quad y \qquad\qquad\qquad\qquad p = 1$$

$$x^2 \quad xy \quad y^2 \qquad\qquad\qquad p = 2$$

$$x^3 \quad x^2y \quad xy^2 \quad y^3 \qquad\qquad p = 3$$

$$x^4 \quad x^3y \quad x^2y^2 \quad xy^3 \quad y^4 \qquad p = 4$$

. . . .

Figure 2.7   Pascal triangle

The described direct derivation of shape functions by the formulation and solution of algebraic equations in terms of nodal coordinates and nodal degrees of freedom is tedious and becomes impractical for higher-order elements. Furthermore, the existence of a solution for these equations (i.e. existence of an inverse for the coefficients matrix in them) is only guaranteed if the elemental interpolations are based on complete polynomials. Important families of useful finite elements do not provide interpolation models that correspond to complete polynomial expansions. Therefore, in practice, indirect methods are employed to derive the shape functions associated with the elements that belong to these families.

A very convenient 'indirect' procedure for the derivation of shape functions in rectangular elements is to use the 'tensor products' of one-dimensional inter- polation functions. This can be readily explained considering the four-node rectangular element shown in Figure 2.8.

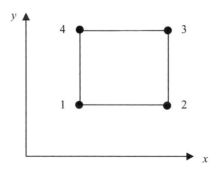

Figure 2.8   Bi-linear rectangular element

The interpolation model in this element is expressed as

$$\tilde{f} = a_1 + a_2 x + a_3 y + a_4 xy \tag{2.15}$$

The polynomial expansion used in this equation does not include all of the terms of a complete quadratic expansion (i.e. six terms corresponding to $p = 2$ in the Pascal triangle) and, therefore, the four-node rectangular element shown in Figure 2.8 is not a quadratic element. The right-hand side of Equation (2.15) can, however, be written as the product of two first-order polynomials in terms of x and y variables as

$$\tilde{f} = (b_1 x + b_2) \cdot (b_3 y + b_4) \tag{2.16}$$

Therefore an obvious procedure for the generation of the shape functions of the element shown in Figure 2.8 is to obtain the products of linear interpolation functions in the x and y directions. The four-noded rectangular element constructed in this way is called a bi-linear element. Higher order members of this family are also readily generated using the tensor products of higher order one-dimensional interpolation functions. For example, the second member of this group is the nine-noded bi-quadratic rectangular element, shown in Figure 2.9, whose shape functions are formulated as the products of quadratic Lagrange polynomials in the x and y directions.

A similar procedure is used to generate 'tensor-product' three-dimensional elements, such as the 27-node tri-quadratic element. The shape functions in two- or three-dimensional tensor product elements are always incomplete polynomials.

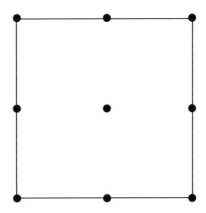

Figure 2.9   Bi-quadratic rectangular element

Analogous to tensor product Lagrange elements, tensor product Hermite elements can also be generated. The rectangular element developed by Bogner *et al.* (1965) is an example of this group. This element is shown in Figure 2.10 and involves a total of 16 degrees of freedom per single variable. The associated shape functions of this element are found as the tensor products of the cubic polynomials in x and y (see Figure 2.5).

Another important group of finite elements whose shape functions are not complete polynomials is the 'serendipity' family. An eight-noded rectangular element which has four corner nodes and four mid-side nodes is an example of this family. Shape functions of serendipity elements cannot be generated by the tensor product of one-dimensional Lagrange interpolation functions (except for the four-node rectangular element which is the same in both families). Hence, these functions are found by an alternative method based on using products of selected polynomials that give desired function variations on element edges (Reddy, 1993).

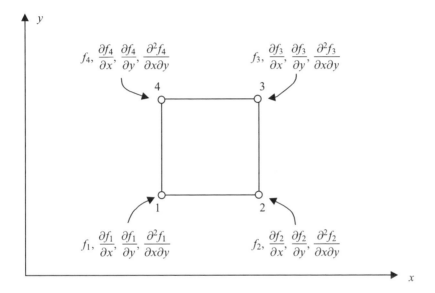

Figure 2.10   A rectangular Hermite element

2.1.3 Non-standard elements

Finite element families described in the previous section are used to obtain standard discretizations in a wide range of different engineering problems. In addition to these families, other element groups that provide specific types of approximations have also been developed. In this section a number of 'non-standard' elements that are widely used to model polymeric flow regimes are described. Taylor–Hood elements are among the earliest examples of this group which were specially designed for the solution of incompressible flow problems. In the Taylor–Hood elements interpolation of pressure is always based on a lower-order polynomial than the polynomials used to interpolate velocity components (Taylor and Hood, 1973). The rectangular element, shown in Figure 2.11, is an example of this family.

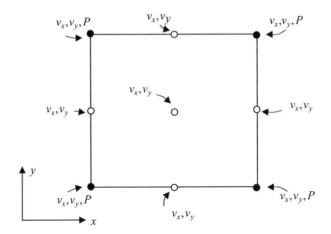

Figure 2.11   Nine node Taylor–Hood element

In this element the velocity and pressure fields are approximated using bi-quadratic and bi-linear shape functions, respectively, this corresponds to a total of 22 degrees of freedom consisting of 18 nodal velocity components (corner, mid-side and centre nodes) and four nodal pressures (corner nodes).

Crouzeix–Raviart elements are another group of finite elements that provide different interpolations for pressure and velocity in a flow domain (Crouzeix and Raviart, 1973). The main characteristic of these elements is to make the pressure on the element boundaries discontinuous. For example, the combination of quadratic shape functions for the approximation of velocity (corresponding to a six-node triangle) with a constant pressure, (given at a single node inside the triangle), can be considered. Another member of this family is the rectangular element shown in Figure 2.12, in which the approximation of velocity is based on bi-quadratic shape functions while pressure is approximated linearly using three internal sampling nodes. This element usually provides a greater flexibility than the Taylor–Hood element, shown in Figure 2.11, in the modelling of incompressible flow problems.

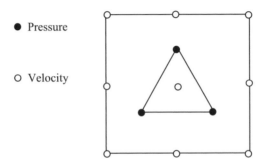

Figure 2.12   An element belonging to the Crouzeix and Raviart family

2.1.4 Local coordinate systems

The global Cartesian framework used so far is not a convenient coordinate system for the generation of function approximations over different elements in a mesh. Elemental shape functions defined in terms of global nodal coordinates will not remain invariant and instead will appear as polynomials of similar degree having different coefficients at each element. This inconvenient situation is readily avoided by using an appropriate local coordinate system to define elemental shape functions. If required, interpolated functions expressed in terms of local coordinates can be transformed to the global coordinate system at a later stage. Shape functions written in terms of local variables will always be the same for a particular finite element no matter what type of global coordinate system is used. Finite element approximation of unknown functions in terms of locally defined shape functions can be written as

$$\tilde{f} = \sum_{i=1}^{p} N_i(x^*) f_i \qquad (2.17)$$

where x^* represents local coordinates and f_i are nodal degrees of freedom. As is shown in Figure 2.13 a local Cartesian coordinate system with its origin located at the centre of the element is the natural choice for rectangular elements.

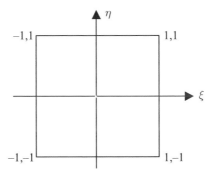

Figure 2.13 Local coordinate system in rectangular elements

Using this coordinate system the shape functions for the first two members of the tensor product Lagrange element family are expressed as

Four-node bi-linear element

$$\begin{cases} N_1 = \frac{1}{2}(1 - \xi) \cdot \frac{1}{2}(1 - \eta) \\ N_2 = \frac{1}{2}(1 + \xi) \cdot \frac{1}{2}(1 - \eta) \\ N_3 = \frac{1}{2}(1 + \xi) \cdot \frac{1}{2}(1 + \eta) \\ N_4 = \frac{1}{2}(1 - \xi) \cdot \frac{1}{2}(1 + \eta) \end{cases} \qquad (2.18)$$

Nine-node bi-quadratic element

$$\begin{cases} N_1 = +\frac{1}{2}\xi(1-\xi)\cdot\frac{1}{2}\eta(1-\eta) \\ \quad N_5 = -(1-\xi^2)\cdot\frac{1}{2}\eta(1-\eta) \\ N_2 = -\frac{1}{2}\xi(1+\xi)\cdot\frac{1}{2}\eta(1-\eta) \\ \quad N_6 = +\frac{1}{2}\xi(1+\xi)\cdot(1-\eta^2) \\ N_3 = +\frac{1}{2}\xi(1+\xi)\cdot\frac{1}{2}\eta(1+\eta) \\ \quad N_7 = +(1-\xi^2)\cdot\frac{1}{2}\eta(1+\eta) \\ N_4 = -\frac{1}{2}\xi(1-\xi)\cdot\frac{1}{2}\eta(1+\eta) \\ \quad N_8 = -\frac{1}{2}\xi(1-\xi)\cdot(1-\eta^2) \\ \quad N_9 = +(1-\xi^2)\cdot(1-\eta^2) \end{cases} \qquad (2.19)$$

the local Cartesian coordinate system used in rectangular elements is not a suitable choice for triangular elements. A natural local coordinate system for the triangular elements can be developed using area coordinates. Consider a triangular element as is shown in Figure 2.14 divided into three sub-areas of A_1, A_2 and A_3. The area coordinates of L_1, L_2 and L_3 for the point P inside this triangle are defined as

$$\begin{cases} L_1 = \dfrac{A_1}{A} \\ L_2 = \dfrac{A_2}{A} \\ L_3 = \dfrac{A_3}{A} \end{cases} \qquad (2.20)$$

where $A = \displaystyle\sum_{i=1}^{3} A_i$ is the total area of the triangle.

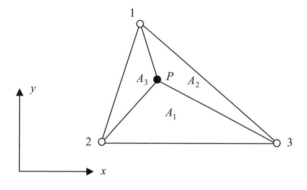

Figure 2.14   Area coordinates in triangular elements

It can be readily shown that $L_i, i = 1,3$ satisfy the requirements for shape functions (as stated in Equation 2.3) associated with the triangular element. The area of a triangle in terms of the Cartesian coordinates of its vertices is written as

$$A = \frac{1}{2} \det \begin{bmatrix} 1 & x_1 & y_1 \\ 1 & x_2 & y_2 \\ 1 & x_3 & y_3 \end{bmatrix} \tag{2.21}$$

Therefore the area coordinates defined by Equation (2.20) in a global Cartesian coordinate system are expressed as

$$L_1 = \frac{A_1}{A} = \frac{\det \begin{bmatrix} 1 & x_2 & y_2 \\ 1 & x_3 & y_3 \\ 1 & x & y \end{bmatrix}}{\det \begin{bmatrix} 1 & x_1 & y_1 \\ 1 & x_2 & y_2 \\ 1 & x_3 & y_3 \end{bmatrix}} \quad \text{etc.} \tag{2.22}$$

Therefore

$$\begin{bmatrix} L_1 \\ L_2 \\ L_3 \end{bmatrix} = \frac{1}{2A} \begin{bmatrix} x_2 y_3 - x_3 y_2 & y_2 - y_3 & x_3 - x_2 \\ x_3 y_1 - x_1 y_3 & y_3 - y_1 & x_1 - x_3 \\ x_1 y_2 - x_2 y_1 & y_1 - y_2 & x_2 - x_1 \end{bmatrix} \begin{bmatrix} 1 \\ x \\ y \end{bmatrix} \tag{2.23}$$

The expansion of Equation (2.23) gives the transformation between the local area coordinates and the global Cartesian system (x, y) for triangular elements. This transformation also confirms that in a global Cartesian coordinate system the shape functions of a linear triangular element should be expressed as Equations (2.12), (2.13) and (2.14). Using the area coordinates the shape functions for the first two members of the triangular finite elements are given as

Three-node linear triangular element

$$\begin{cases} N_1 = L_1 \\ N_2 = L_2 \\ N_3 = L_3 \end{cases} \tag{2.24}$$

Six-node quadratic triangular element

$$\begin{cases} N_1 = 2L_1^2 - L_1 \\ N_4 = 4L_1L_2 \\ N_2 = 2L_2^2 - L_2 \\ N_5 = 4L_2L_3 \\ N_3 = 2L_3^2 - L_3 \\ N_6 = 4L_3L_1 \end{cases} \qquad (2.25)$$

2.1.5 Order of continuity of finite elements

A general requirement in most finite element discretizations is to maintain the compatibility of field variables (or functions) across the boundaries of the neighbouring elements. Finite elements that generate uniquely defined function approximations along their sides (boundaries) satisfy this condition. For example, in a mesh consisting of three-node triangular elements with nodes at its vertices, linear interpolation used to derive the element shape functions gives a unique variation for functions along each side of the element. Therefore, field variables or functions on the nodes of this element are uniquely defined. This example can be contrasted with a three node triangular element in which the nodes (i.e. sampling points for interpolation) are located at mid-points of the triangle, as is shown in Figure 2.15. Clearly it will not be possible to obtain unique linear variations for functions along the sides of the triangular element shown in the figure.

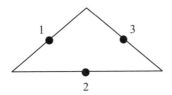

Figure 2.15 A nodal arrangement that cannot provide inter-element compatibility of functions

Finite elements that maintain inter-element compatibility of functions are called 'conforming elements'. Finite elements that do not have this property are referred to as the 'non-conforming elements'. Under certain conditions non-conforming elements can lead to accurate solutions and are more advantageous to use.

The order of continuity of a conforming finite element that only ensures the compatibility of functions across its boundaries is said to be C^0. Finite elements that ensure the inter-element compatibility of functions and their derivatives provide a higher order of continuity than C^0. For example, the Hermite element shown in Figure 2.4 which guarantees the compatibility of function values and slopes at its ends is C^1 continuous. According to this definition, the non-conforming triangular element shown in Figure 2.15 is said to be a C^{-1} continuous element.

The order of continuity and the degree of highest order complete polynomial obtainable in an elemental interpolation are used to identify various finite elements. For example, the three-node linear triangle, shown in Figure 2.6, and the four-node bi-linear rectangle, shown in Figure 2.8, are both referred to as P^1C^0 elements. Similarly, according to this convention, the nine-node bi-quadratic rectangle, shown in Figure 2.9, is said to be a P^2C^0 element and so on.

2.1.6 Convergence

All numerical computations inevitably involve round-off errors. This error increases as the number of calculations in the solution procedure is increased. Therefore, in practice, successive mesh refinements that increase the number of finite element calculations do not necessarily lead to more accurate solutions. However, one may assume a theoretical situation where the rounding error is eliminated. In this case successive reduction in size of elements in the mesh should improve the accuracy of the finite element solution. Therefore, using a P^pC^n element with sufficient orders of interpolation and continuity, at the limit (i.e. when element dimensions tend to zero), an exact solution should be obtained. This has been shown to be true for linear elliptic problems (Strang and Fix, 1973) where an optimal convergence is achieved if the following conditions are satisfied:

- aspect ratio of quadrilateral elements should be small,

- internal angles of elements should not be near $0°$ or $180°$,

- the exact solution should be sufficiently smooth and must not include singularities,

- problem domain should be convex, and

- elemental calculations (i.e. evaluation of integrals etc.) must be sufficiently accurate.

Theoretical analysis of convergence in non-linear problems is incomplete and in most instances does not yield clear results. Conclusions drawn from the analyses of linear elliptic problems, however, provide basic guidelines for solving non-linear or non-elliptic equations.

2.1.7 Irregular and curved elements – isoparametric mapping

Flexibility to cope with irregular domain geometry in a straightforward and systematic manner is one of the most important characteristics of the finite element method. Irregular domains that do not include any curved boundary sections can be accurately discretized using triangular elements. In most engineering processes, however, the elimination of discretization error requires the use of finite elements which themselves have curved sides. It is obvious that randomly shaped curved elements cannot be developed in an ad hoc manner and a general approach that is applicable in all situations must be sought. The required generalization is obtained using a two step procedure as follows:

- a regular element called the 'master element' is selected and a local finite element approximation based on the shape functions of this element is established, and

- the master element is mapped into the global coordinates to generate the required distorted elements.

A graphical representation of this process is shown in Figure 2.16.

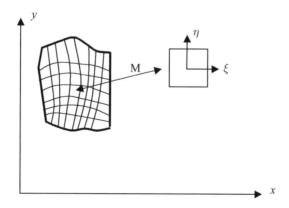

Figure 2.16   Mapping between a master element and elements in a global mesh

In the figure operation (M) represents a one-to-one transformation between the local and global coordinate systems. This in general can be shown as

$$M \Rightarrow \begin{cases} x = x(\xi, \eta) \\ y = y(\xi, \eta) \end{cases} \tag{2.26}$$

The one-to-one transformation between the global and local coordinate systems can be established using a variety of techniques (Zienkiewicz and Morgan,

1983). The most general method is a form of 'parametric mapping' in which the transformation functions, $x(\xi,\eta)$ and $y(\xi,\eta)$ in Equation (2.26), are polynomials based on the element shape functions. Three different forms of this technique have been developed:

- Subparametric transformations: shape functions used in the mapping functions are lower-order polynomials than the shape functions used to obtain finite element approximation of functions.

- Superparametric transformations: shape functions used in the mapping functions are higher-order polynomials than the shape functions used to obtain finite element approximation of functions.

- Isoparametric transformations: shape functions used in the mapping functions are identical to the shape functions used to obtain finite element approximation of functions.

Isoparametric mapping is the most commonly used form of the described parametric transformation. Figure 2.17 shows a schematic example of isoparametric transformation between an irregular element and its corresponding regular (master) element. Shape functions along the sides of the master element shown in this example are linear in ξ and η and consequently they can only generate irregular elements with straight sides. In contrast the master element shown in Figure 2.18 can be mapped into a global element with curved sides.

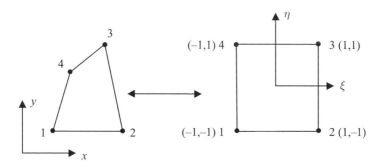

Figure 2.17   Isoparametric mapping of an irregular quadrilateral element with straight sides

In general, elements with curved sides can only be generated using quadratic or higher-order master elements.

Isoparametric transformation functions between a global coordinate system and local coordinates are, in general, written as

$$
\begin{cases}
x = \displaystyle\sum_{i=1}^{p} N_i(\xi, \eta) x_i \\[2mm]
y = \displaystyle\sum_{i=1}^{p} N_i(\xi, \eta) y_i
\end{cases}
\tag{2.27}
$$

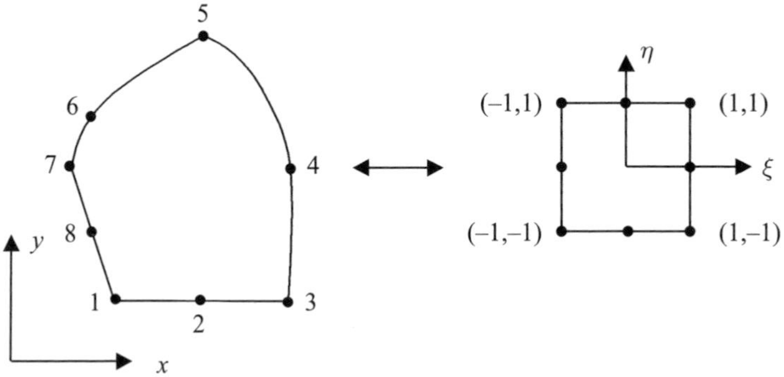

Figure 2.18 Isoparametric mapping of an irregular quadrilateral element with curved sides

where x_i and y_i are the nodal coordinates in the global system. The shape functions in Equation (2.27) are given in terms of local variables defined by the natural coordinate system in the master element. Isoparametric mapping can also be used to generate triangular elements with curved sides. As already explained, the local variables in triangular elements are given as area coordinates and hence isoparametric mapping functions for triangular elements are expressed as

$$
\begin{cases}
x = \displaystyle\sum_{i=1}^{p} L_i x_i \\[2mm]
y = \displaystyle\sum_{i=1}^{p} L_i y_i
\end{cases}
\tag{2.28}
$$

The most convenient coordinate system for a triangular master element is based on a 'natural' system similar to the one shown in Figure 2.19, where $L_1 = 1 - \xi - \eta$, $L_2 = \xi$ and $L_3 = \eta$.

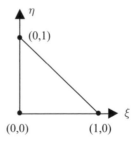

Figure 2.19 Local natural coordinates in a master triangular element

In addition to the stated condition of one-to-one correspondence between local and global coordinates the transformation must preserve the geometrical conformity and continuity of the mesh in a way that no gaps or overlapping can occur between the elements. To satisfy this, irregular elements in a mesh should be generated by mappings from a master element that has an appropriate order of continuity. For example, C^0 continuous distorted elements should be generated from C^0 continuous parent elements.

In conjunction with the use of isoparametric elements it is necessary to express the derivatives of nodal functions in terms of local coordinates. This is a straightforward procedure for elements with C^0 continuity and can be described as follows: Using the chain rule for differentiation of functions of multiple variables, the derivative of a function in terms of local variables (ξ, η) can be expressed as

$$\frac{\partial \phi(\xi, \eta)}{\partial \xi} = \frac{\partial \phi(\xi, \eta)}{\partial x} \frac{\partial x}{\partial \xi} + \frac{\partial \phi(\xi, \eta)}{\partial y} \frac{\partial y}{\partial \xi} \tag{2.29}$$

and

$$\frac{\partial \phi(\xi, \eta)}{\partial \eta} = \frac{\partial \phi(\xi, \eta)}{\partial x} \frac{\partial x}{\partial \eta} + \frac{\partial \phi(\xi, \eta)}{\partial y} \frac{\partial y}{\partial \eta} \tag{2.30}$$

Using matrix notations

$$\begin{bmatrix} \dfrac{\partial \phi(\xi, \eta)}{\partial \xi} \\[2ex] \dfrac{\partial \phi(\xi, \eta)}{\partial \eta} \end{bmatrix} = \begin{bmatrix} \dfrac{\partial x}{\partial \xi} & \dfrac{\partial y}{\partial \xi} \\[2ex] \dfrac{\partial x}{\partial \eta} & \dfrac{\partial y}{\partial \eta} \end{bmatrix} \begin{bmatrix} \dfrac{\partial \phi(\xi, \eta)}{\partial x} \\[2ex] \dfrac{\partial \phi(\xi, \eta)}{\partial y} \end{bmatrix} \tag{2.31}$$

or

$$\begin{bmatrix} \dfrac{\partial \phi}{\partial \xi} \\[2ex] \dfrac{\partial \phi}{\partial \eta} \end{bmatrix} = J \begin{bmatrix} \dfrac{\partial \phi}{\partial x} \\[2ex] \dfrac{\partial \phi}{\partial y} \end{bmatrix} \tag{2.32}$$

where J is the Jacobian of coordinate transformations. Therefore

$$\begin{bmatrix} \dfrac{\partial \phi}{\partial x} \\[2ex] \dfrac{\partial \phi}{\partial y} \end{bmatrix} = J^{-1} \begin{bmatrix} \dfrac{\partial \phi}{\partial \xi} \\[2ex] \dfrac{\partial \phi}{\partial \eta} \end{bmatrix} \tag{2.33}$$

Global derivatives of functions can now be related to the locally defined finite element approximation, given by Equation (2.17), as

$$
\begin{bmatrix} \dfrac{\partial \tilde{f}}{\partial x} \\[2ex] \dfrac{\partial \tilde{f}}{\partial y} \end{bmatrix} = \boldsymbol{J}^{-1} \begin{bmatrix} \dfrac{\partial \sum\limits_{i=1}^{p} N_i(\xi,\eta)f_i}{\partial \xi} \\[4ex] \dfrac{\partial \sum\limits_{i=1}^{p} N_i(\xi,\eta)f_i}{\partial \eta} \end{bmatrix}
\tag{2.34}
$$

Obviously the described transformation depends on the existence of an inverse for the Jacobian matrix (i.e. det $\boldsymbol{J}$ must always be non-zero).

Differentiation of locally defined shape functions appearing in Equation (2.34) is a trivial matter, in addition, in isoparametric elements members of the Jacobian matrix are given in terms of locally defined derivatives and known global coordinates of the nodes (Equation 2.27). Consequently, computation of the inverse of the Jacobian matrix shown in Equation (2.34) is usually straightforward.

It should be emphasized at this point that the isoparametric mapping of regular elements into curved shapes inevitably generates a degree of approximation. The magnitude of the error in such approximations directly depends on the degree of the irregularity of the elements being mapped. In general, mappings involving badly distorted elements in coarse meshes should be avoided. In extreme situations the sign of the Jacobian changes during the transformation and an illogical element that folds upon itself is generated.

In certain types of finite element computations the application of isoparametric mapping may require transformation of second-order as well as the first-order derivatives. Isoparametric transformation of second (or higher)-order derivatives is not straightforward and requires lengthy algebraic manipulations. Details of a convenient procedure for the isoparametric transformation of second-order derivatives are given by Petera *et al.* (1993).

Isoparametric mapping removes the geometrical inflexibility of rectangular elements and therefore they can be used to solve many types of practical problems. For example, the isoparametric C^1 continuous rectangular Hermite element provides useful discretizations in the solution of viscoelastic flow problems.

2.1.8 Numerical integration

The finite element solution of differential equations requires function integration over element domains. Evaluation of integrals over elemental domains by analytical methods can be tedious and impractical and is not attempted in general. Furthermore, in the majority of problems isoparametric mapping is used to generate meshes involving irregular and curved elements and hence the analytical evaluation of elemental integrals is practically impossible. Therefore, in finite element computations integrals given over elemental domains are found by numerical integration (quadrature) techniques. A commonly used quadrature method is the application of the Gauss–Legendre formula for the evaluation of definite integrals between limits of −1 and +1. This procedure is summarized using the following example:

Consider the integration of a function $f(x_1, x_2)$ over a quadrilateral element in a finite element mesh expressed as

$$I = \int_{\alpha_1}^{\beta_1} \int_{\alpha_2}^{\beta_2} f(x_1, x_2) dx_1 dx_2 \tag{2.35}$$

where x_1 and x_2 are the global coordinates and α_1 etc. are constants. The transformation between the global and local domains gives

$$I = \int_{-1}^{1} \int_{-1}^{1} F(\xi, \eta) \det \boldsymbol{J}^e d\xi d\eta \tag{2.36}$$

where the limits of integration are defined by the local coordinates and the integration measure is transformed as

$$dx_1 dx_2 = \det \boldsymbol{J}^e d\xi d\eta \tag{2.37}$$

After algebraic manipulations we can write

$$I = \int_{-1}^{1} \int_{-1}^{1} G(\xi, \eta) d\xi d\eta \tag{2.38}$$

Using the Gauss–Legendre quadrature, I is found as

$$I = \int_{-1}^{1} \int_{-1}^{1} G(\xi, \eta) d\xi d\eta \approx \sum_{I=1}^{M} \sum_{J=1}^{N} G(\xi_I, \eta_J) W_I W_J \tag{2.39}$$

where ξ_I and η_J are the quadrature point coordinates, W_I and W_J are the corresponding weight factors and M and N are the number of quadrature points in each summation. The number of quadrature points in these summations depends on the order of the polynomial function in the integral. In one-dimensional problems this quadrature yields an exact result for a polynomial of degree $2n-1$ (or less) using n points. In finite element computations, integrands in elemental equations are based on shape functions, which are low-order poly-nomials, therefore the number of required quadrature points is low (usually $n =$ 2 or 3). Table 2.1 shows the coordinates of quadrature points and their associated weighting factors for $M = 3$ in the Gauss–Legendre formula.

Table 2.1

$M = N$	W_I	$\xi_I,\ \eta_I$
1	2	0
2	1	$\pm\ 0.5773502692$
3	0.5555555556	$\pm\ 0.7745966692$
	0.8888888889	0

As already mentioned, the local coordinate system in triangular elements is defined in terms of area coordinates. Therefore, in these elements the integration limits will also be given in terms of area coordinates precluding their evaluation by the described Gauss–Legendre quadrature. For details of quadrature techniques, sampling points and weighting factors for triangular elements see Zienkiewicz and Taylor (1994).

2.1.9   Mesh refinement – h- and p-versions of the finite element method

The standard technique for improving the accuracy of finite element approxi-mations is to refine the computational grid in order to use a denser mesh con-sisting of smaller size elements. This also provides a practical method for testing the convergence in the solution of non-linear problems through the comparison of the results obtained on successively refined meshes. In the 'h-version' of the finite element method the element selected for the domain discretization remains unchanged while the number and size of the elements vary with each level of mesh refinement. Alternatively, the accuracy of the finite element discretizations can be enhanced using higher-order elements whilst the basic mesh design is kept constant. For example, after obtaining a solution for a problem on a mesh consisting of bi-linear elements another solution is generated via bi-quadratic elements while keeping the number, size and shape of the elements in the mesh unchanged. In this case the number of the nodes and, consequently, the node-to-element ratio in the mesh will increase and a better accuracy will be obtained. This approach is commonly called the 'p-version' of the finite element method.

The family of 'hierarchical' elements are specifically designed to minimize the computational cost of repeated computations in the 'p-version' of the finite element method (Zienkiewicz and Taylor, 1994). Successive approximations based on hierarchical elements utilize the derivations of a lower step to generate the solution for a higher-order approximation. This can significantly reduce the cost of repeated computations in p-version finite element simulations. However, hierarchical elements lack the flexibility of ordinary elements and their application is restricted to specific problems (Bathe, 1996).

2.2 NUMERICAL SOLUTION OF DIFFERENTIAL EQUATIONS BY THE WEIGHTED RESIDUAL METHOD

The weighted residual method provides a flexible mathematical framework for the construction of a variety of numerical solution schemes for the differential equations arising in engineering problems. In particular, as is shown in the following section, its application in conjunction with the finite element discretizations yields powerful solution algorithms for field problems. To outline this technique we consider a steady-state boundary value problem represented by the following mathematical model

$$\begin{cases} \Im[u(x)] = g \text{ in } \Omega \\ \text{subject to} \\ u(x) = v \text{ on } \Gamma \end{cases} \tag{2.40}$$

where $\Im$ is a linear differential operator, $u(x)$ is the unknown variable (function of independent spatial variables), g is a source/sink term, Ω is a sufficiently smooth closed domain surrounded by a continuous boundary Γ and v is the specified value of $u(x)$ at Γ. In the absence of an exact analytical solution for Equation (2.40) we seek to represent the field variable $u(x)$ approximately as

$$u(x) \approx \tilde{u}(x) = v + \sum_{i=1}^{m} \alpha_i \Phi_i(x) \tag{2.41}$$

where α_i ($i = 1,m$) are a set of coefficients (constants) and Φ_i ($i = 1,m$) represent a set of geometrical functions called *basis functions*. To satisfy the boundary conditions the defined approximation should have the following property $\tilde{u}|_\Gamma = u|_\Gamma = v$ whatever the values chosen for the constants. Therefore the basis functions should be selected in a way that $\Phi_i|_\Gamma = 0$ for all i. Accuracy and convergence of the defined approximation will depend on the selected basis functions and as a fundamental rule these functions should be chosen in a way that the approximation becomes more accurate as m increases. Substitution from Equation (2.41) into Equation (2.40) gives

$$\Im[\tilde{u}(x)] - g = R_\Omega \tag{2.42}$$

where $R_\Omega \neq 0$ is the residual which will inevitably appear through the insertion of an approximation instead of an exact solution for the field variable into the differential equation. Equation (2.42) is written as

$$\Im\left[v + \sum_{i=1}^{m} \alpha_i \Phi_i(x)\right] - g = R_\Omega \tag{2.43}$$

The residual, R_Ω, is a function of position in Ω. The weighted residual method is based on the elimination of this residual, in some overall manner, over the entire domain. To achieve this the residual is weighted by an appropriate number of position dependent functions and a summation is carried out. This is written as

$$\int_\Omega W_j R_\Omega d\Omega = 0 \quad j = 1, 2, 3, \ldots m \tag{2.44}$$

where W_j are linearly independent weight functions and $d\Omega$ is an appropriate integration measure. Substitution of R_Ω from Equation (2.42) into Equation (2.44) gives

$$\int_\Omega W_j \left\{ \Im\left[v + \sum_{i=1}^{m} \alpha_i \Phi_i(x)\right] - g \right\} d\Omega = 0 \quad j = 1, 2, 3, \ldots m \tag{2.45}$$

Equation (2.45) represents the 'weighted residual statement' of the original differential equation. Theoretically, this equation provides a system of m simultaneous linear equations, with coefficients α_i, $i = 1, \ldots m$, as unknowns, that can be solved to obtain the unknown coefficients in Equation (2.41). Therefore, the required approximation (i.e. the discrete solution) of the field variable becomes determined.

Despite the simplicity of the outlined weighted residual method, its application to the solution of practical problems is not straightforward. The main difficulty arises from the lack of any systematic procedure that can be used to select appropriate basis and weight functions in a problem. The combination of finite element approximation procedures with weighted residual methods resolves this problem. This is explained briefly in the forthcoming section.

2.2.1   Weighted residual statements in the context of finite element discretizations

As already discussed, variations of a field unknown within a finite element is approximated by the shape functions. Therefore finite element discretization provides a natural method for the construction of piecewise approximations for the unknown functions in problems formulated in a global domain. This is readily ascertained considering the mathematical model represented by Equation (2.40). After the discretization of Ω into a mesh of finite elements weighted residual statement of Equation (2.40), within the space of a finite element Ω_e, is written as

$$\int_{\Omega_e} W_j \left\{ \left[\Im \sum_{i=1}^{p} N_i(\pmb{x}) u_i \right] - g \right\} d\Omega_e = 0 \quad j = 1, \ldots, p \tag{2.46}$$

where

$$\sum_{e=1}^{E} \Omega_e = \Omega, \quad E = \text{total number of elements}$$

In Equation (2.46) the unknown function u is approximated using the shape functions of Ω_e in the usual manner. It is important to note at this stage that the elemental discretization used to obtain Equation (2.46) should guarantee the existence of a finite integral for all of the terms in the weighted residual statement. Therefore the selection of appropriate elements for a problem primarily depends on the nature of the differential operator $\Im$ in the original model equations. For example, in the solution of second order differential equations it appears that elemental discretizations that provide inter-element continuity at least up to the first-order derivatives should be used. This restriction can, however, be eased to a large degree by the application of Green's theorem (integration by parts) to the second-order derivatives to reduce the order of differentiations to one. The weakening of the continuity requirement achieved by this operation is an important aspect of many practical schemes used in engineering problems. Further discussion of the formulation of 'weak statements' in the finite element modelling of polymer flow problems can be found in Chapter 3.

For each W_j, Equation (2.46) generates a corresponding equation and collectively these equations can be shown using matrix notation as

$$[A]\{u_i\} = \{b\} \tag{2.47}$$

where $[A]$ and $\{b\}$ are called the elemental 'stiffness matrix' and 'load vector', respectively. Repeated application of the method to all elements in the finite element mesh and subsequent assembly of elemental stiffness matrices and load vectors over the common nodes yields a global set of algebraic equations in terms of the nodal unknowns. After the prescription of the boundary conditions the assembled global set of equations becomes determinate and can be solved. An important property of the finite element method is that it always produces a sparse banded set of global equations. This is an inherent property of the method and should hence be used to achieve computing economy.

2.2.2   The standard Galerkin method

In the standard Galerkin method (also called the Bubnov–Galerkin method) weight functions in the weighted residual statements are selected to be identical to the basis functions (Zienkiewicz and Morgan, 1983). Therefore the standard Galerkin representation of Equation (2.46) is given as

$$\int_{\Omega_e} N_j \left\{ \left[\Im \sum_{i=1}^{p} N_i(x)u_i \right] - g \right\} d\Omega_e = 0 \quad j = 1, \ldots, p \tag{2.48}$$

In conjunction with Equation (2.48) the prescribed boundary conditions at the boundaries of the solution domain should also be satisfied.

The simplicity gained by choosing identical weight and shape functions has made the standard Galerkin method the most widely used technique in the finite element solution of differential equations. Because of the centrality of this technique in the development of practical schemes for polymer flow problems, the entire procedure of the Galerkin finite element solution of a field problem is further elucidated in the following worked example.

2.2.2*   Galerkin finite element procedure – a worked example

As an illustrative example we consider the Galerkin finite element solution of the following differential equation in domain Ω, as shown in Figure 2.20.

$$\begin{cases} \dfrac{d^2 T}{dx^2} + T = 0 \quad \text{in} \quad \Omega \\[2mm] \text{subject to:} \\ T_A = 0, T_B = 1 \end{cases} \tag{2.49}$$

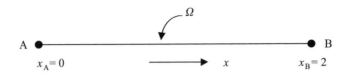

Figure 2.20   One-dimensional problem domain

Step 1: discretization of the problem domain

The domain Ω is discretized into a mesh of five unequal size linear finite elements, as is shown in Figure 2.21.

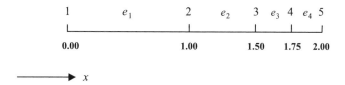

Figure 2.21   Discretization of the problem domain

Step 2: approximation using shape functions

Within the space of finite elements the unknown function is approximated using shape functions corresponding to the two-noded (linear) Lagrange elements as

$$\tilde{T} = \sum_{i=1}^{2} N_i(x) T_i \tag{2.50}$$

where $N_i(x)$, $i = 1,2$ are the shape functions and T_i, $i = 1,2$ are the nodal degrees of freedom (i.e. nodal unknowns).

Step 3: Galerkin-weighted residual statement

The residual obtained via the insertion of $\tilde{T}$ into the differential equation is weighted and integrated over each element as

$$\int_{\Omega_e} w \left(\frac{d^2 \tilde{T}}{dx^2} + \tilde{T} \right) dx = 0 \tag{2.51}$$

where w is a weighting function. In the standard Galerkin method the selected weight functions are identical to the shape functions and hence Equation (2.51) is written as

$$\int_{\Omega_e} N_j \left[\frac{d^2 \sum_{i=1}^{2} N_i(x) T_i}{dx^2} + \sum_{i=1}^{2} N_i(x) T_i \right] dx = 0 \tag{2.52}$$

Step 4: integration by parts (Green's theorem)

At this stage the formulated Galerkin-weighted residual Equation (2.52) contains second-order derivatives. Therefore C^0 elements cannot generate an acceptable solution for this equation (using C^0 elements the first derivative of the shape functions will be discontinuous across element boundaries and the integral of their second derivative will tend to infinity). To solve this difficulty the second-order derivative in Equation (2.52) is integrated by parts to obtain the 'weak' form of the weighted residual statement as

$$-\int_{\Omega_e}\left(\frac{\mathrm{d}\sum N_i(x)T_i}{\mathrm{d}x}\cdot\frac{\mathrm{d}N_j}{\mathrm{d}x}\right)\mathrm{d}x+\int_{\Omega_e}N_j\sum N_i(x)T_i\mathrm{d}x+N_j\frac{\mathrm{d}\sum N_i(x)T_i}{\mathrm{d}x}\Big|_{\Gamma_e}=0 \qquad (2.53)$$

where Γ_e represents an element boundary (for simplicity summation limits are not written).

Step 5: formulation of the elemental stiffness equations

The weight function used in the Galerkin formulation can be identical to either of the shape functions of a two-node linear element, therefore, for each weight function an equation corresponding to the weak statement (2.53) is derived

$$\begin{cases}-\int_{\Omega_e}\frac{\mathrm{d}}{\mathrm{d}x}(N_IT_I+N_{II}T_{II})\frac{\mathrm{d}N_I}{\mathrm{d}x}\mathrm{d}x+\int_{\Omega_e}(N_IT_I+N_{II}T_{II})N_I\mathrm{d}x+N_I\phi|_{\Gamma_e}=0\\[2mm]-\int_{\Omega_e}\frac{\mathrm{d}}{\mathrm{d}x}(N_IT_I+N_{II}T_{II})\frac{\mathrm{d}N_{II}}{\mathrm{d}x}\mathrm{d}x+\int_{\Omega_e}(N_IT_I+N_{II}T_{II})N_{II}\mathrm{d}x+N_{II}\phi|_{\Gamma_e}=0\end{cases} \qquad (2.54)$$

where ϕ represents the boundary line term. Using matrix notation, Equation (2.54) is written as

$$\begin{bmatrix}-\int_{\Omega_e}\left(\dfrac{\mathrm{d}N_I}{\mathrm{d}x}\dfrac{\mathrm{d}N_I}{\mathrm{d}x}-N_IN_I\right)\mathrm{d}x & -\int_{\Omega_e}\left(\dfrac{\mathrm{d}N_{II}}{\mathrm{d}x}\dfrac{\mathrm{d}N_I}{\mathrm{d}x}-N_{II}N_I\right)\mathrm{d}x\\[4mm]-\int_{\Omega_e}\left(\dfrac{\mathrm{d}N_I}{\mathrm{d}x}\dfrac{\mathrm{d}N_{II}}{\mathrm{d}x}-N_IN_{II}\right)\mathrm{d}x & -\int_{\Omega_e}\left(\dfrac{\mathrm{d}N_{II}}{\mathrm{d}x}\dfrac{\mathrm{d}N_{II}}{\mathrm{d}x}-N_{II}N_{II}\right)\mathrm{d}x\end{bmatrix}\begin{Bmatrix}T_I\\T_{II}\end{Bmatrix}=\begin{Bmatrix}-N_I\phi|_{\Gamma_e}\\-N_{II}\phi|_{\Gamma_e}\end{Bmatrix} \qquad (2.55)$$

Although the elemental stiffness Equation (2.55) has a common form for all of the elements in the mesh, its utilization based on the shape functions defined in the global coordinate system is not convenient. This is readily ascertained considering that shape functions defined in the global system have different coefficients in each element. For example

$$N_I = 1 - x \quad \text{and} \quad N_{II} = x \quad \text{in} \quad e_1$$

and

$$N_I = \frac{1.5 - x}{0.5} \quad \text{and} \quad N_{II} = \frac{x - 1}{0.5} \quad \text{in} \quad e_2$$

Furthermore, in a global system limits of definite integrals in the coefficient matrix will be different for each element. This difficulty is readily resolved using a local coordinate system (shown as χ) to define the elemental shape functions as

$$\begin{cases} N_I = \dfrac{\ell_e - \chi}{\ell_e} \\[3mm] N_{II} = \dfrac{\chi}{\ell_e} \end{cases} \quad \text{and} \quad \begin{cases} \dfrac{dN_I}{d\chi} = \dfrac{-1}{\ell_e} \\[3mm] \dfrac{dN_{II}}{d\chi} = \dfrac{1}{\ell_e} \end{cases} \tag{2.56}$$

where ℓ_e is the element length. Therefore Equation (2.55) is written as

$$\begin{bmatrix} -\displaystyle\int_0^{\ell_e}\left(\dfrac{dN_I}{d\chi}\dfrac{dN_I}{d\chi} - N_I N_I\right)d\chi & -\displaystyle\int_0^{\ell_e}\left(\dfrac{dN_{II}}{d\chi}\dfrac{dN_I}{d\chi} - N_{II}N_I\right)d\chi \\[4mm] -\displaystyle\int_0^{\ell_e}\left(\dfrac{dN_I}{d\chi}\dfrac{dN_{II}}{d\chi} - N_I N_{II}\right)d\chi & -\displaystyle\int_0^{\ell_e}\left(\dfrac{dN_{II}}{d\chi}\dfrac{dN_{II}}{d\chi} - N_{II}N_{II}\right)d\chi \end{bmatrix} \begin{Bmatrix} T_I \\ T_{II} \end{Bmatrix} = \begin{Bmatrix} -N_I\phi|_{\Gamma_e} \\ -N_{II}\phi|_{\Gamma_e} \end{Bmatrix} \tag{2.57}$$

Substitution for the shape functions from Equation (2.56) into Equation (2.57) gives

$$\frac{1}{(\ell_e)^2}\begin{bmatrix} \displaystyle\int_0^{\ell_e}(-1 + \ell_e^2 - 2\ell_e\chi + \chi^2)d\chi & \displaystyle\int_0^{\ell_e}(1 + \ell_e\chi - \chi^2)d\chi \\[4mm] \displaystyle\int_0^{\ell_e}(1 + \ell_e\chi - \chi^2)d\chi & \displaystyle\int_0^{\ell_e}(-1 + \chi^2)d\chi \end{bmatrix}\begin{Bmatrix} T_I \\ T_{II} \end{Bmatrix} = \begin{Bmatrix} -\dfrac{\ell_e - \chi}{\ell_e}\phi\big|_0^{\ell_e} \\[4mm] -\dfrac{\chi}{\ell_e}\phi\big|_0^{\ell_e} \end{Bmatrix} \tag{2.58}$$

After the evaluation of the definite integrals in the coefficient matrix and the boundary line terms in the right-hand side, Equation (2.58) gives

$$\begin{bmatrix} \dfrac{\ell_e}{3} - \dfrac{1}{\ell_e} & \dfrac{1}{\ell_e} + \dfrac{\ell_e}{6} \\[4mm] \dfrac{1}{\ell_e} + \dfrac{\ell_e}{6} & \dfrac{\ell_e}{3} - \dfrac{1}{\ell_e} \end{bmatrix}\begin{Bmatrix} T_I \\ T_{II} \end{Bmatrix} = \begin{Bmatrix} q_I \\ -q_{II} \end{Bmatrix} \tag{2.59}$$

Therefore for e_1, $\ell_e = 1$

$$\begin{bmatrix} -2/3 & 7/6 \\ 7/6 & -2/3 \end{bmatrix} \begin{Bmatrix} T_1 \\ T_2 \end{Bmatrix} = \begin{Bmatrix} q_1 \\ -q_2 \end{Bmatrix} \tag{2.60}$$

For e_2, $\ell_e = 1/2$

$$\begin{bmatrix} -11/6 & 25/12 \\ 25/12 & -11/6 \end{bmatrix} \begin{Bmatrix} T_2 \\ T_3 \end{Bmatrix} = \begin{Bmatrix} q_2 \\ -q_3 \end{Bmatrix} \tag{2.61}$$

For e_3, $\ell_e = 1/4$

$$\begin{bmatrix} -47/12 & 97/24 \\ 97/24 & -47/12 \end{bmatrix} \begin{Bmatrix} T_3 \\ T_4 \end{Bmatrix} = \begin{Bmatrix} q_3 \\ -q_4 \end{Bmatrix} \tag{2.62}$$

For e_4, $\ell_e = 1/4$

$$\begin{bmatrix} -47/12 & 97/24 \\ 97/24 & -47/12 \end{bmatrix} \begin{Bmatrix} T_4 \\ T_5 \end{Bmatrix} = \begin{Bmatrix} q_4 \\ -q_5 \end{Bmatrix} \tag{2.63}$$

Step 6: assembly of the elemental stiffness equations into a global system of algebraic equations

Elemental stiffness equations are assembled over their common nodes to yield

$$\begin{bmatrix} -2/3 & 7/6 & 0 & 0 & 0 \\ 7/6 & -15/6 & 25/12 & 0 & 0 \\ 0 & 25/12 & -69/12 & 97/24 & 0 \\ 0 & 0 & 97/24 & -47/6 & 97/24 \\ 0 & 0 & 0 & 97/24 & -47/12 \end{bmatrix} \begin{Bmatrix} T_1 \\ T_2 \\ T_3 \\ T_4 \\ T_5 \end{Bmatrix} = \begin{Bmatrix} q_1 \\ 0 \\ 0 \\ 0 \\ -q_5 \end{Bmatrix} \tag{2.64}$$

Note that in equation system (2.64) the coefficients matrix is symmetric, sparse (i.e. a significant number of its members are zero) and banded. The symmetry of the coefficients matrix in the global finite element equations is not guaranteed for all applications (in particular, in most fluid flow problems this matrix will not be symmetric). However, the finite element method always yields sparse and banded sets of equations. This property should be utilized to minimize computing costs in complex problems.

Step 7: imposition of the boundary conditions

Prescribed values of the unknown function at the boundaries of Ω (i.e. $T_1 = 0$, $T_5 = 1$) are inserted into the system of algebraic equations (2.64) and redundant equations corresponding to the boundary nodes eliminated from the set. After algebraic manipulations the following set of equations is obtained

$$
\begin{bmatrix} -15/6 & 25/12 & 0 \\ 25/12 & -69/12 & 97/24 \\ 0 & 97/24 & -47/6 \end{bmatrix} \begin{Bmatrix} T_2 \\ T_3 \\ T_4 \end{Bmatrix} = \begin{Bmatrix} 0 \\ 0 \\ -97/24 \end{Bmatrix}
\tag{2.65}
$$

Step 8: solution of the algebraic equations

Equation set (2.65) is a determinate system and its solution gives

$$
\begin{cases} T_2 = 0.9011 \\ T_3 = 1.0813 \\ T_4 = 1.0738 \end{cases}
$$

At this point by repeating the above solution on a grid constructed by the division of the domain given in Figure 2.20 into 10 elements of equal size the effect of mesh refinement on the accuracy of the standard Galerkin procedure can be demonstrated. The general elemental stiffness equation is identical to Equation (2.59) which after the insertion of an equal element length of $\ell_e = 0.2$ gives the stiffness coefficient matrix for all of the elements as

$$
\begin{bmatrix} -74/15 & 151/30 \\ 151/30 & -74/15 \end{bmatrix}
$$

Therefore the assembled global stiffness matrix in this case is written as

$$
\begin{bmatrix}
-74/15 & 151/30 & 0 & 0 & 0 & 0 & 0 & 0 & 0 & 0 & 0 \\
151/30 & -148/15 & 151/30 & 0 & 0 & 0 & 0 & 0 & 0 & 0 & 0 \\
0 & 151/30 & -148/15 & 151/30 & 0 & 0 & 0 & 0 & 0 & 0 & 0 \\
0 & 0 & 151/30 & -148/15 & 151/30 & 0 & 0 & 0 & 0 & 0 & 0 \\
0 & 0 & 0 & 151/30 & -148/15 & 151/30 & 0 & 0 & 0 & 0 & 0 \\
0 & 0 & 0 & 0 & 151/30 & -148/15 & 151/30 & 0 & 0 & 0 & 0 \\
0 & 0 & 0 & 0 & 0 & 151/30 & -148/15 & 151/30 & 0 & 0 & 0 \\
0 & 0 & 0 & 0 & 0 & 0 & 151/30 & -148/15 & 151/30 & 0 & 0 \\
0 & 0 & 0 & 0 & 0 & 0 & 0 & 151/30 & -148/15 & 151/30 & 0 \\
0 & 0 & 0 & 0 & 0 & 0 & 0 & 0 & 151/30 & -148/15 & 151/30 \\
0 & 0 & 0 & 0 & 0 & 0 & 0 & 0 & 0 & 151/30 & -74/15
\end{bmatrix}
$$

After the insertion of the boundary conditions the solution of the system of algebraic equations in this case gives the required nodal values of T (i.e. T_2 to T_{10}) as

$$\left\{ \begin{array}{l} T_2 = 0.2178 \\ T_3 = 0.4269 \\ T_4 = 0.6191 \\ T_5 = 0.7867 \\ T_6 = 0.9230 \\ T_7 = 1.0227 \\ T_8 = 1.0817 \\ T_9 = 1.0977 \\ T_{10} = 1.0709 \end{array} \right.$$

The analytical solution of Equation (2.49) is

$$T = 1.09975 \sin(x) \qquad\qquad (2.66)$$

Figure 2.22 shows the comparison of the analytical solution with the Galerkin finite element (FE) results obtained using the 4 and 10 element grids.

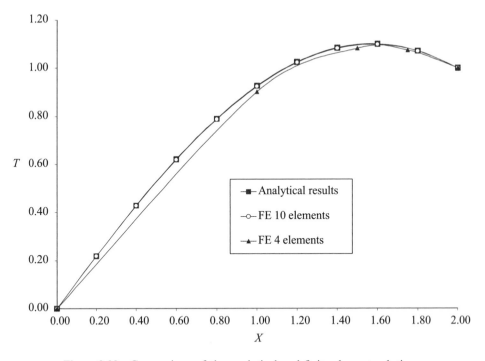

Figure 2.22 Comparison of the analytical and finite element solutions

In the finite element solution of engineering problems the global set of equations obtained after the assembly of elemental contributions will be very large (usually consisting of several thousand algebraic equations). They may also be ill-conditioned (Gerald and Wheatley, 1984). Therefore the solution of the global system of equations is regarded as one of the most important steps in the finite element modelling of realistic problems. Various 'direct' elimination techniques, such as 'frontal solution' or 'LU decomposition', and 'iterative' procedures, such as the 'preconditioned conjugate gradient method', are used as *equation solvers* in the finite element programs. Computing economy, speed and the required accuracy of the solutions are the most important factors that should be taken into account in selecting solver routines for finite element programming. Descriptions and full listings of computer codes based on these techniques can be found in the literature (e.g. see Forsythe and Meler, 1967; Hood, 1976; Hinton and Owen, 1977; Irons and Ahmad, 1980). Further explanations about the solution of the systems of algebraic equations arising in finite element computations are given in Chapter 6.

Coordinate transformation between local and global systems – mapping

Isoparametric mapping described in Section 1.7 for generating curved and distorted elements is not, in general, relevant to one-dimensional problems. However, the problem solved in this section provides a simple example for the illustration of important aspects of this procedure. Consider a 'master' element as is shown in Figure 2.23. The shape functions associated with this element are

$$
\begin{cases}
N_I = \dfrac{1-\xi}{2} \\[2mm]
N_{II} = \dfrac{1+\xi}{2}
\end{cases}
\tag{2.67}
$$

$$
\frac{dN_I}{d\xi} = -\frac{1}{2} \quad \text{and} \quad \frac{dN_{II}}{d\xi} = \frac{1}{2}
$$

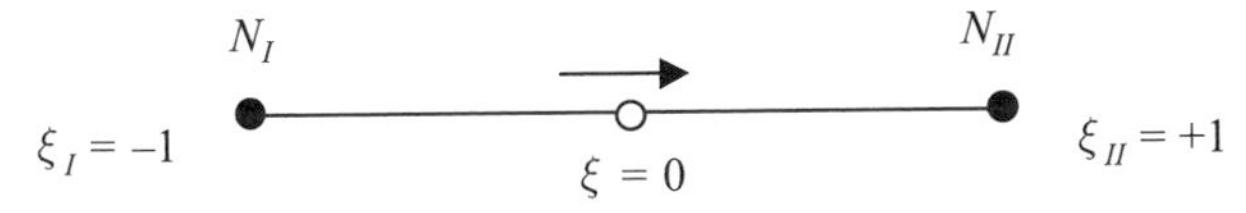

Figure 2.23   Isoparametric master element

Therefore the approximate form of the unknown function within this element is written as

$$
\tilde{T} = N_I T_I + N_{II} T_{II} = \frac{1-\xi}{2} T_I + \frac{1+\xi}{2} T_{II}
\tag{2.68}
$$

In order to establish an isoparametric mapping between the master element shown in Figure 2.23 and the elements in the global domain (Figure 2.20) we use the elemental shape functions to formulate a transformation function as

$$x = N_I X_I + N_{II} X_{II} = \frac{1-\xi}{2} X_I + \frac{1+\xi}{2} X_{II} \tag{2.69}$$

hence

$$\frac{\mathrm{d}x}{\mathrm{d}\xi} = -\frac{1}{2} X_I + \frac{1}{2} X_{II} = \frac{\ell_e}{2} \qquad \rightarrow \mathrm{d}x = \frac{\ell_e}{2} \mathrm{d}\xi \tag{2.70}$$

where X_I and X_{II} and are the nodal coordinate values in the global system and ℓ_e represents the length of a linear element. Thus the derivatives of the shape functions in the global system are found as

$$\begin{cases} \dfrac{\mathrm{d}N_I}{\mathrm{d}x} = \dfrac{1}{\dfrac{\mathrm{d}x}{\mathrm{d}\xi}} \cdot \dfrac{\mathrm{d}N_I}{\mathrm{d}\xi} = \dfrac{2}{\ell_e} \times -\dfrac{1}{2} = -\dfrac{1}{\ell_e} \\[4mm] \dfrac{\mathrm{d}N_{II}}{\mathrm{d}x} = \dfrac{1}{\dfrac{\mathrm{d}x}{\mathrm{d}\xi}} \cdot \dfrac{\mathrm{d}N_{II}}{\mathrm{d}\xi} = \dfrac{2}{\ell_e} \times \dfrac{1}{2} = +\dfrac{1}{\ell_e} \end{cases} \tag{2.71}$$

Note that in the one-dimensional problem illustrated here the Jacobian of coordinate transformation is simply expressed as $\mathrm{d}x/\mathrm{d}\xi$ and therefore

$$J^{-1} = \frac{1}{\dfrac{\mathrm{d}x}{\mathrm{d}\xi}} \tag{2.72}$$

After the substitution for $\tilde{T}$ from Equation (2.68), $\mathrm{d}x$ from Equation (2.70) and global derivatives of shape functions from Equation (2.71) into the elemental stiffness equation (2.55) we obtain, for the equation corresponding to N_I

$$\int\limits_{-1}^{+1} \left(-\frac{1}{\ell_e} T_I + \frac{1}{\ell_e} T_{II} \right) \times \frac{1}{\ell_e} \times \frac{\ell_e}{2} \mathrm{d}\xi + \int\limits_{-1}^{+1} (N_I T_I + N_{II} T_I) N_I \frac{\ell_e}{2} \mathrm{d}\xi + N_I|_{-1}^{+1} q_I = 0 \tag{2.73}$$

or

$$\int\limits_{-1}^{+1} \left(-\frac{1}{\ell_e} T_I + \frac{1}{\ell_e} T_{II} \right) \frac{1}{2} \mathrm{d}\xi + \int\limits_{-1}^{+1} \left[\frac{1-\xi}{2} T_I + \frac{1+\xi}{2} T_{II} \right] \frac{1-\xi}{2} \frac{\ell_e}{2} \mathrm{d}\xi + \frac{1-\xi}{2} |_{-1}^{+1} q_I = 0 \tag{2.74}$$

and

$$\frac{1}{2\ell_e} \int_{-1}^{+1} (-T_I + T_{II})\mathrm{d}\xi + \frac{\ell_e}{8} \int_{-1}^{+1} [(1 - \xi)^2 T_I + (1 - \xi^2) T_{II}]\mathrm{d}\xi - q_I = 0 \qquad (2.75)$$

Similarly for the second equation in set (2.55) using N_{II} as the weight function

$$-\frac{1}{2\ell_e} \int_{-1}^{+1} (-T_I + T_{II})\mathrm{d}\xi + \frac{\ell_e}{8} \int_{-1}^{+1} [(1 - \xi^2) T_I + (1 + \xi)^2 T_{II}]\mathrm{d}\xi - q_{II} = 0 \quad (2.76)$$

And the elemental stiffness equation is written as

$$\begin{bmatrix} \int_{-1}^{+1} \left(-\frac{1}{2\ell_e} + \frac{\ell_e}{8}(1 - \xi)^2\right)\mathrm{d}\xi & \int_{-1}^{+1} \left(\frac{1}{2\ell_e} + \frac{\ell_e}{8}(1 - \xi^2)\right)\mathrm{d}\xi \\ \int_{-1}^{+1} \left(\frac{1}{2\ell_e} + \frac{\ell_e}{8}(1 - \xi^2)\right)\mathrm{d}\xi & \int_{-1}^{+1} \left(-\frac{1}{2\ell_e} + \frac{\ell_e}{8}(1 + \xi)^2\right)\mathrm{d}\xi \end{bmatrix} \begin{Bmatrix} T_I \\ T_{II} \end{Bmatrix} = \begin{Bmatrix} q_I \\ q_{II} \end{Bmatrix}$$

$$(2.77)$$

and

$$\begin{bmatrix} \dfrac{\ell_e}{3} - \dfrac{1}{\ell_e} & \dfrac{1}{\ell_e} + \dfrac{\ell_e}{6} \\ \dfrac{1}{\ell_e} + \dfrac{\ell_e}{6} & \dfrac{\ell_e}{3} - \dfrac{1}{\ell_e} \end{bmatrix} \begin{Bmatrix} T_I \\ T_{II} \end{Bmatrix} = \begin{Bmatrix} q_I \\ -q_{II} \end{Bmatrix} \qquad (2.78)$$

As it can be seen equation sets of (2.78) and (2.59) are identical.

Note that the definite integrals in the members of the elemental stiffness matrix in Equation (2.77) are given, uniformly, between the limits of –1 and +1. This provides an important facility for the evaluation of the members of the elemental matrices in finite element computations by a systematic numerical integration procedure (see Section 1.8).

2.2.3 Streamline upwind Petrov–Galerkin method

The standard Galerkin technique provides a flexible and powerful method for the solution of problems in areas such as solid mechanics and heat conduction where the model equations are of elliptic or parabolic type. It can also be used to develop robust schemes for the solution of the governing equations of continuity and motion in creeping (very low Reynolds number) incompressible flow regimes encountered in many types of polymer processing systems. However, partial differential equations describing convection dominated phenomena, such as high Peclet number heat transfer or viscoelastic constitutive behaviour are of hyperbolic type and cannot be solved by this method. Application of the Galerkin method to these problems gives unstable and oscillatory results unless proper procedures are adopted to stabilize the solution.

Development of weighted residual finite element schemes that can yield stable solutions for hyperbolic partial differential equations has been the subject of a considerable amount of research. The most successful outcome of these attempts is the development of the streamline upwinding technique by Brooks and Hughes (1982). The basic concept in the streamline upwinding is to modify the weighting function in the Galerkin scheme as

$$W_I = N_I + \alpha N_{I,i} \frac{\bar{v}_i}{|v|^2} \tag{2.79}$$

where N_I and $N_{I,i}$ are a shape function and its derivative, respectively, v is the velocity vector and α is a coefficient called the upwinding parameter (in Equation (2.79) summation over the repeated index is assumed). Determination of the appropriate upwinding parameter in multi-dimensional problems is not straightforward. In general, the analysis described by Brooks and Hughes (1982) for the derivation of the optimum upwinding parameters in one-dimensional problems is used, heuristically, to define the upwinding parameters in multi-dimensional problems (Pittman and Nakazawa, 1984). In practice, numerical tests based on trial and error may be needed to find the appropriate level of upwinding required in a problem.

In the earlier versions of the streamline upwinding scheme the modified weight function was only applied to the convection terms (i.e. first-order derivatives in the hyperbolic equations) while all other terms were weighted in the usual manner. This is called 'selective' or 'inconsistent' upwinding. Selective upwinding can be interpreted as the introduction of an 'artificial diffusion' in addition to the physical diffusion to the weighted residual statement of the differential equation. This improves the stability of the scheme but the accuracy of the solution declines.

Streamline upwinding schemes can also be compared with the Petrov–Galerkin methods. Petrov–Galerkin techniques are a class of Galerkin-weighted residual methods in which the weight functions are not identical to the basis functions (i.e. shape functions in the finite element context). These schemes offer greater flexibility than the standard Galerkin method in dealing with problems such as the solution of hyperbolic partial differential equations. Therefore it is natural to consider streamline upwind Galerkin schemes as a type of Petrov–Galerkin method. However, in contrast to the selective upwinding, in streamline upwind Petrov–Galerkin schemes the modified weight function (here shown as Equation (2.79)) is applied consistently to all terms in the weighted residual statement.

To illustrate the basic concepts described in this section we consider the following worked example.

2.2.3* Application of upwinding – a worked example

We consider the finite element solution of the following differential equation in domain Ω, as shown in Figure 2.24 (an identical domain to problem 2.2.2* is used)

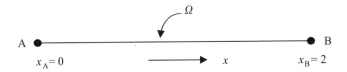

Figure 2.24 The problem domain

$$
\begin{cases}
\dfrac{\mathrm{d}^2 T}{\mathrm{d}x^2} + a\dfrac{\mathrm{d}T}{\mathrm{d}x} + T = 0 \\[2mm]
\text{subject to:} \\
T_A = 0, \, T_B = 1
\end{cases} \tag{2.80}
$$

Let us first consider the standard Galerkin solution of Equation (2.80) obtained using the previously described steps.

Following the discretization of the solution domain Ω (i.e. line AB) into two-node Lagrange elements, and representation of T as $T = \sum N_i(x)T_i)$ in terms of shape functions $N_i(x)$, $i = 1,2$ within the space of a finite element Ω_e, the elemental Galerkin-weighted residual statement of the differential equation is written as

$$
\int_{\Omega_e} N_J \left(\frac{\mathrm{d}^2 \sum N_i(x)T_i}{\mathrm{d}x^2} + a\frac{\mathrm{d}\sum N_i(x)T_i}{\mathrm{d}x} + \sum N_i(x)T_i \right) \mathrm{d}x = 0 \tag{2.81}
$$

After the application of Green's theorem to the second order term in Equation (2.81) we get the weak form of the residual statement as

$$
-\int_{\Omega_e} \left(\frac{\mathrm{d}\sum N_i(x)T_i}{\mathrm{d}x} \cdot \frac{\mathrm{d}N_j}{\mathrm{d}x} \right) \mathrm{d}x + a\int_{\Omega_e} N_J \frac{\mathrm{d}\sum N_i(x)T_i}{\mathrm{d}x} \mathrm{d}x + \int_{\Omega_e} N_j \sum N_i(x)T_i \mathrm{d}x +
$$

$$
N_j \frac{\mathrm{d}\sum N_i(x)T_i}{\mathrm{d}x}\Big|_{\Gamma_e} = 0 \tag{2.82}
$$

where Γ_e represents an element boundary. Based on Equation (2.82) the elemental stiffness equation is formulated as

$$
\left\{
\begin{aligned}
&-\int_{\Omega_e} \frac{\mathrm{d}}{\mathrm{d}x}(N_I T_I + N_{II} T_{II})\frac{\mathrm{d}N_I}{\mathrm{d}x}\mathrm{d}x + a\int_{\Omega_e}\frac{\mathrm{d}}{\mathrm{d}x}(N_I T_I + N_{II} T_{II})N_I \mathrm{d}x + \\
&\qquad\qquad\qquad \int_{\Omega_e}(N_I T_I + N_{II} T_{II})N_I \mathrm{d}x + N_I\phi|_{\Gamma_e} = 0 \\[8pt]
&-\int_{\Omega_e} \frac{\mathrm{d}}{\mathrm{d}x}(N_I T_I + N_{II} T_{II})\frac{\mathrm{d}N_{II}}{\mathrm{d}x}\mathrm{d}x + a\int_{\Omega_e}\frac{\mathrm{d}}{\mathrm{d}x}(N_I T_I + N_{II} T_{II})N_{II} \mathrm{d}x + \\
&\qquad\qquad\qquad \int_{\Omega_e}(N_I T_I + N_{II} T_{II})N_{II} \mathrm{d}x + N_{II}\phi|_{\Gamma_e} = 0
\end{aligned}
\right.
\tag{2.83}
$$

where ϕ represents the boundary line term. Using matrix notation Equation (2.83) is written as

$$
\left[
\begin{array}{cc}
-\int_{\Omega_e}\left(\dfrac{\mathrm{d}N_I}{\mathrm{d}x}\dfrac{\mathrm{d}N_I}{\mathrm{d}x} - a\dfrac{\mathrm{d}N_I}{\mathrm{d}x}N_I - N_I N_I\right)\mathrm{d}x & -\int_{\Omega_e}\left(\dfrac{\mathrm{d}N_{II}}{\mathrm{d}x}\dfrac{\mathrm{d}N_I}{\mathrm{d}x} - a\dfrac{\mathrm{d}N_{II}}{\mathrm{d}x}N_I - N_{II}N_I\right)\mathrm{d}x \\[10pt]
-\int_{\Omega_e}\left(\dfrac{\mathrm{d}N_I}{\mathrm{d}x}\dfrac{\mathrm{d}N_{II}}{\mathrm{d}x} - a\dfrac{\mathrm{d}N_I}{\mathrm{d}x}N_{II} - N_I N_{II}\right)\mathrm{d}x & -\int_{\Omega_e}\left(\dfrac{\mathrm{d}N_{II}}{\mathrm{d}x}\dfrac{\mathrm{d}N_{II}}{\mathrm{d}x} - a\dfrac{\mathrm{d}N_{II}}{\mathrm{d}x}N_{II} - N_{II}N_{II}\right)\mathrm{d}x
\end{array}
\right]
\left\{
\begin{array}{c}
T_I \\ T_{II}
\end{array}
\right\}
$$

$$
= \left\{
\begin{array}{c}
-N_I\phi|_{\Gamma_e} \\ -N_{II}\phi|_{\Gamma_e}
\end{array}
\right\}
\tag{2.84}
$$

Note that in contrast to the example shown in Section 2.2.2* the element stiffness equation obtained for this problem is not symmetric. After the substitution for the shape functions and algebraic manipulations

$$
\frac{-1}{\ell_e^2}\left(
\begin{array}{cc}
\int_0^{\ell_e}\left[1 + a(\ell_e - x) - (\ell_e - x)^2\right]\mathrm{d}x & \int_0^{\ell_e}\left[-1 - a(\ell_e - x) - x(\ell_e - x)\right]\mathrm{d}x \\[10pt]
\int_0^{\ell_e}\left[-1 + ax - x(\ell_e - x)\right]\mathrm{d}x & \int_0^{\ell_e}\left(1 - ax - x^2\right)\mathrm{d}x
\end{array}
\right)
\left(
\begin{array}{c}
T_I \\ T_{II}
\end{array}
\right)
= \left(
\begin{array}{c}
q_I \\ -q_{II}
\end{array}
\right)
\tag{2.85}
$$

After the evaluation of the integrals in the terms of the coefficient matrix, we have

$$
\left(\begin{array}{cc} \left(\dfrac{-1}{\ell_e} - \dfrac{a}{2} + \dfrac{\ell_e}{3}\right) & \left(\dfrac{1}{\ell_e} + \dfrac{a}{2} + \dfrac{\ell_e}{6}\right) \\[2ex] \left(\dfrac{+1}{\ell_e} - \dfrac{a}{2} + \dfrac{\ell_e}{6}\right) & \left(-\dfrac{1}{\ell_e} + \dfrac{a}{2} + \dfrac{\ell_e}{3}\right) \end{array}\right) \begin{pmatrix} T_I \\[2ex] T_{II} \end{pmatrix} = \begin{pmatrix} q_I \\[2ex] -q_{II} \end{pmatrix} \tag{2.86}
$$

Choosing a domain discretization based on 10 elements of equal size ($\ell_e = 0.2$), we have

$$
\left(\begin{array}{cc} (-4.934 - a/2) & (5.033 + a/2) \\[2ex] (5.033 - a/2) & (-4.934 + a/2) \end{array}\right) \begin{pmatrix} T_I \\[2ex] T_{II} \end{pmatrix} = \begin{pmatrix} q_I \\[2ex] -q_{II} \end{pmatrix} \tag{2.87}
$$

After the assembly and insertion of the boundary conditions the following set of global stiffness equations is derived

$$
\begin{bmatrix}
d & c & 0 & 0 & 0 & 0 & 0 & 0 & 0 \\
s & d & c & 0 & 0 & 0 & 0 & 0 & 0 \\
0 & s & d & c & 0 & 0 & 0 & 0 & 0 \\
0 & 0 & s & d & c & 0 & 0 & 0 & 0 \\
0 & 0 & 0 & s & d & c & 0 & 0 & 0 \\
0 & 0 & 0 & 0 & s & d & c & 0 & 0 \\
0 & 0 & 0 & 0 & 0 & s & d & c & 0 \\
0 & 0 & 0 & 0 & 0 & 0 & s & d & c \\
0 & 0 & 0 & 0 & 0 & 0 & 0 & s & d
\end{bmatrix}
\begin{bmatrix}
T_2 \\ T_3 \\ T_4 \\ T_5 \\ T_6 \\ T_7 \\ T_8 \\ T_9 \\ T_{10}
\end{bmatrix}
=
\begin{bmatrix}
0 \\ 0 \\ 0 \\ 0 \\ 0 \\ 0 \\ 0 \\ 0 \\ -c
\end{bmatrix}
\tag{2.88}
$$

where $d = -9.868$, $c = 5.033 + (a/2)$ and $s = 5.033 - (a/2)$. We can now obtain the solution for this problem and compare it with the analytical result.

The analytical solution of Equation (2.80) with the given boundary conditions for $a = 1$ is

$$
T = 2.754\, e^{-\frac{1}{2}x} \sin\frac{\sqrt{3}}{2}x \tag{2.89}
$$

The comparison between the finite element and analytical solutions for a relatively small value of $a = 1$ is shown in Figure 2.25. As can be seen the standard Galerkin method has yielded an accurate and stable solution for the differential Equation (2.80). The accuracy of this solution is expected to improve even further with mesh refinement. As Figure 2.26 shows using $a = 10$ a stable result can still be obtained, however using the present mesh of 10 elements, for larger values of this coefficient the numerical solution produced by the standard

Galerkin method becomes unstable and useless. It can also be seen that these oscillations become more intensified as a becomes larger (note that the factor affecting the stability is the magnitude of a and oscillatory solutions will also result using large negative coefficients).

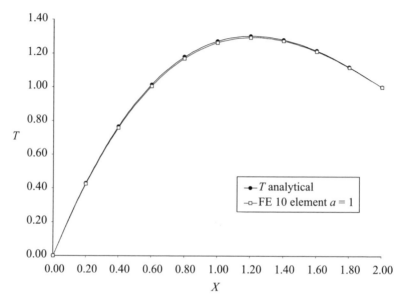

Figure 2.25 Comparison of the analytical and finite element results for a low Peclet number problem

The first order derivative in Equation (2.80) corresponds to the convection in a field problem and the examples shown in Figure 2.26 illustrates the inability of the standard Galerkin method to produce meaningful results for convection-dominated equations. As described in the previous section to resolve this difficulty, in the solution of hyperbolic (convection-dominated) equations, upwinding or Petrov–Galerkin methods are employed. To demonstrate the application of upwinding we consider the case where only the weight function applied to the first-order derivative in the weak variational statement of the problem, represented by Equation (2.82), is modified.

The weighted residual statement corresponding to Equation (2.80) is hence written as

$$\int_{\Omega_e} N_J \left(\frac{\mathrm{d}^2 \sum N_i(x)T_i}{\mathrm{d}x^2} + \sum N_i(x)T_i \right) \mathrm{d}x + N_J^* \left(a \frac{\mathrm{d} \sum N_i(x)T_i}{\mathrm{d}x} \right) \mathrm{d}x = 0 \quad (2.90)$$

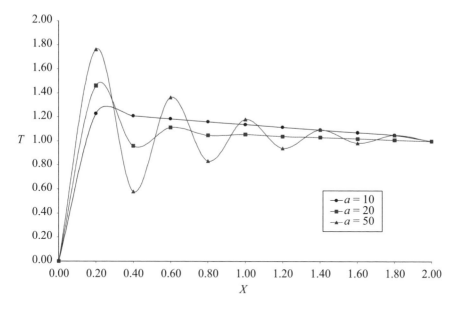

Figure 2.26 Comparison of the finite element solution of low and high Peclet number problems

Integration by parts (Green's theorem) of the second order term in Equation (2.90) gives the weak form of the problem as

$$-\int_{\Omega_e}\left(\frac{d\sum N_i(x)T_i}{dx}\frac{dN_j}{dx}\right)dx + \int_{\Omega_e} N_j \sum N_i(x)T_i dx + N_j \frac{d\sum N_i(x)T_i}{dx}\bigg|_{\Gamma_e}$$

$$+a\int_{\Omega_e} N_J^* \frac{d\sum N_i(x)T_i}{dx}dx = 0 \qquad (2.91)$$

Using two-noded Lagrangian elements the shape functions are given as

$$N_I = \frac{\ell_e - x}{\ell_e} \quad \text{and} \quad N_{II} = \frac{x}{\ell_e}$$

therefore

$$\frac{dN_I}{dx} = -\frac{1}{\ell_e} \quad \text{and} \quad \frac{dN_{II}}{dx} = \frac{1}{\ell_e}$$

In the simple one-dimensional example considered here the upwinded weight function found using Equation (2.89) is reduced to $W = N + \beta(dN/dx)$. Therefore, the modified weight functions applied to the first order derivative term in Equation (2.91) can be written as

$$
\begin{cases}
N_I^* = N_I + \beta \dfrac{dN_I}{dx} = \dfrac{\ell_e - x}{\ell_e} - \dfrac{\beta}{\ell_e} = \dfrac{\ell_e - x - \beta}{\ell_e} \\[4mm]
N_{II}^* = N_{II} + \beta \dfrac{dN_{II}}{dx} = \dfrac{x}{\ell_e} + \dfrac{\beta}{\ell_e} = \dfrac{x + \beta}{\ell_e}
\end{cases}
\tag{2.92}
$$

The general elemental stiffness equation can thus be written as

$$
\left[
\begin{array}{cc}
-\displaystyle\int_{\Omega_e} \left(\dfrac{dN_I}{dx}\dfrac{dN_I}{dx} - a\dfrac{dN_I}{dx}N_I^* - N_I N_I \right) dx & -\displaystyle\int_{\Omega_e} \left(\dfrac{dN_{II}}{dx}\dfrac{dN_I}{dx} - a\dfrac{dN_{II}}{dx}N_I^* - N_{II} N_I \right) dx \\[4mm]
-\displaystyle\int_{\Omega_e} \left(\dfrac{dN_I}{dx}\dfrac{dN_{II}}{dx} - a\dfrac{dN_I}{dx}N_{II}^* - N_I N_{II} \right) dx & -\displaystyle\int_{\Omega_e} \left(\dfrac{dN_{II}}{dx}\dfrac{dN_{II}}{dx} - a\dfrac{dN_{II}}{dx}N_{II}^* - N_{II} N_{II} \right) dx
\end{array}
\right]
\begin{Bmatrix} T_I \\ T_{II} \end{Bmatrix}
$$

$$
= \begin{Bmatrix} -N_I \phi |_{\Gamma_e} \\ -N_{II} \phi |_{\Gamma_e} \end{Bmatrix}
\tag{2.93}
$$

Substitution of the shape functions gives

$$
\dfrac{-1}{\ell_e^2}
\left(
\begin{array}{cc}
\displaystyle\int_0^{\ell_e} \left[1 + a(\ell_e - x - \beta) - (\ell_e - x)^2 \right] dx & \displaystyle\int_0^{\ell_e} [-1 - a(\ell_e - x - \beta) - x(\ell_e - x)] dx \\[4mm]
\displaystyle\int_0^{\ell_e} [-1 + a(x + \beta) - x(\ell_e - x)] dx & \displaystyle\int_0^{\ell_e} \left(1 - a(x + \beta) - x^2 \right) dx
\end{array}
\right)
\begin{pmatrix} T_I \\ T_{II} \end{pmatrix}
=
\begin{pmatrix} q_I \\ -q_{II} \end{pmatrix}
\tag{2.94}
$$

After integration

$$
\left[
\begin{array}{cc}
\left(-\dfrac{1}{\ell_e} + \dfrac{a\beta}{\ell_e} - \dfrac{a}{2} + \dfrac{\ell_e}{3} \right) & \left(\dfrac{1}{\ell_e} + \dfrac{a}{2} - \dfrac{a\beta}{\ell_e} + \dfrac{\ell_e}{6} \right) \\[4mm]
\left(\dfrac{1}{\ell_e} - \dfrac{a}{2} - \dfrac{a\beta}{\ell_e} + \dfrac{\ell_e}{6} \right) & \left(\dfrac{-1}{\ell_e} + \dfrac{a\beta}{\ell_e} + \dfrac{a}{2} + \dfrac{\ell_e}{3} \right)
\end{array}
\right]
\begin{pmatrix} T_I \\ T_{II} \end{pmatrix}
=
\begin{pmatrix} q_I \\ -q_{II} \end{pmatrix}
\tag{2.95}
$$

Choosing a mesh of 10 elements of equal size we have

$$
\left[
\begin{array}{cc}
(-4.933 + 5.0a\beta - a/2) & (5.033 + a/2 - 5.0a\beta) \\[2mm]
(5.033 - a/2 - 5.0a\beta) & (-4.933 + 5.0a\beta + a/2)
\end{array}
\right]
\begin{pmatrix} T_I \\ T_{II} \end{pmatrix}
=
\begin{pmatrix} q_I \\ -q_{II} \end{pmatrix}
\tag{2.96}
$$

As an example we consider the solution of Equation (2.80) with a value of $a = 50$, in this case the general form of the elemental stiffness equation is written as

$$\begin{bmatrix} (-29.933 + 250.0\beta) & (30.033 - 250.0\beta) \\ (-19.967 - 250.0\beta) & (20.067 + 250.0\beta) \end{bmatrix} \begin{pmatrix} T_I \\ T_{II} \end{pmatrix} = \begin{pmatrix} q_I \\ -q_{II} \end{pmatrix} \tag{2.97}$$

After the assembly of elemental equations into a global set and imposition of the boundary conditions the final solution of the original differential equation with respect to various values of upwinding parameter β can be found. The analytical solution of Equation (2.80) with $a = 50$ is found as

$$T = 1.0408(e^{-0.02x} - e^{-49.98x}) \tag{2.98}$$

The finite element results obtained for various values of β are compared with the analytical solution in Figure 2.27. As can be seen using a value of $\beta = 0.5$ a stable numerical solution is obtained. However, this solution is over-damped and inaccurate. Therefore the main problem is to find a value of upwinding parameter that eliminates oscillations without generating over-damped results. To illustrate this concept let us consider the following convection–diffusion equation

$$\alpha \frac{d^2 T}{dx^2} - v \frac{dT}{dx} = 0$$

in Ω subject to:
$$\begin{cases} x = 0, T = T_0 \\ x = L, T = T_L \end{cases} \tag{2.99}$$

where the constants α and v represent diffusivity and velocity, respectively. Note that the selection of a 'hard' downstream boundary condition is intentional to make the numerical solution prone to oscillations, with use of a 'soft' boundary condition such as $x = L$, $(dT/dx) = 0$ stable results may be obtained without difficulty. At the most basic level upwinding can be viewed as the introduction of a certain amount of artificial diffusivity into the problem so that instead of Equation (2.99) the following equation is solved

$$(\alpha + \gamma) \frac{d^2 T}{dx^2} - v \frac{dT}{dx} = 0$$

in Ω subject to:
$$\begin{cases} x = 0, T = T_0 \\ x = L, T = T_L \end{cases} \tag{2.100}$$

Theoretical analysis by Hughes and Brooks (1979) has shown that using a value of

$$\gamma = \frac{vh}{2} \left[\coth \left(\frac{P_e}{2} \right) - \frac{2}{P_e} \right] \tag{2.101}$$

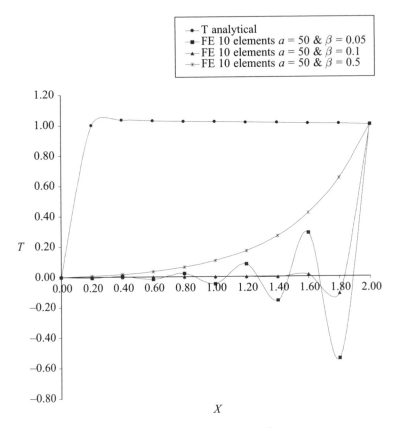

Figure 2.27   Effect of upwinding

where h is the length of individual linear elements in the computational mesh and $P_e = (vh/\alpha)$ is the mesh Peclet number (a dimensionless number representative of the relative dominance of convection over diffusion in energy transport), the standard Galerkin solution of Equation (2.100) will be exactly the same as the analytical solution of Equation (2.99). Extending this approach to the solution of Equation (2.80), with the coefficient of the first-order derivative term $a = 50$, the mesh Peclet number corresponding to a discretization of the problem domain into 10 linear elements of equal size is calculated. Insertion of this number into Equation (2.101) yields the artificial diffusivity, $\gamma = 4.0$. After addition of this value to the coefficient of the second-order term the standard Galerkin solution of Equation (2.80) can be found. However, in the streamline upwinding method the calculated value of γ is taken to be the upwinding parameter, i.e. α in Equation (2.79). This gives the 'appropriate value' of the upwinding coefficient in the modified weight function $W = N + \beta(dN/dx)$ as $\beta = -0.08$. Repeating the procedure outlined through Equation (2.90) to (2.97) the streamline upwind solution of Equation (2.80) can now be found. As comparison of this solution with the analytical result, given in Figure

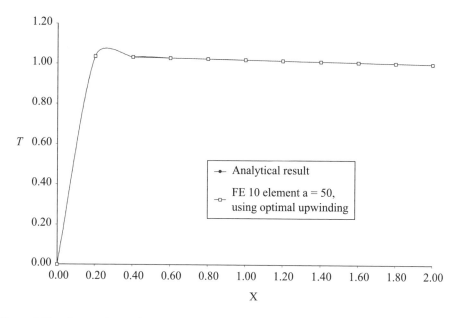

Figure 2.28 Comparison of the analytical solution with the finite element result obtained using optimal upwinding

2.28, shows the numerical solution is not affected by any cross-wind diffusion and hence is very accurate; hence optimal upwinding has generated a 'superconvergent' solution which coincides with the analytical result. Hughes and Brooks (1979) extended the approach to multi-dimensional problems and developed the 'streamline upwinding' technique in which the modified weight functions are adjusted to introduce the necessary dissipation only along the streamlines in a flow field and avoid excessive over-damping by 'cross-wind' diffusion. In multi-dimensional problems, however, it is impossible to develop a theoretical analysis that can yield an exact value for the required artificial diffusivity analogous to γ given by Equation (2.101) for the one-dimensional case. Based on the concept of streamline-upwinding modified weight functions such as

$$N^* = N + \gamma_c \frac{|h_1 v_1 + h_2 v_2|}{2|\bar{v}|^2}\left(v_1 \frac{\partial N}{\partial x_1} + v_2 \frac{\partial N}{\partial x_2}\right) \qquad (2.102)$$

where $0 < \gamma_c \geq 1$ is a constant, v_1 and v_2 are velocity components and h_1 and h_2 are characteristic element lengths, can be used to obtain stable solutions for two-dimensional problems. In general, however, because of the impossibility of selection of an optimum value for γ_c that can guarantee the elimination of all spurious cross-wind diffusion and indeed the uncertainty of definition of element length in a two-dimensional domain the numerical solution is expected to involve some degree of upwinding error.

2.2.4   Least-squares finite element method

The standard least-squares approach provides an alternative to the Galerkin method in the development of finite element solution schemes for differential equations. However, it can also be shown to belong to the class of weighted residual techniques (Zienkiewicz and Morgan, 1983). In the least-squares finite element method the sum of the squares of the residuals, generated via the substitution of the unknown functions by finite element approximations, is formed and subsequently minimized to obtain the working equations of the scheme. The procedure can be illustrated by the following example, consider

$$\Im(u) - g = 0 \text{ on } \Omega \tag{2.103}$$

For simplicity we assume that $\Im = (d^2/dx^2) + \alpha(d/dx)$, where α is a constant. The basic steps in the least squares scheme are:

(a)   formulation of a functional using the squares of the residual obtained by the substitution of the unknown as

$$u \approx \bar{u} = \sum_{1}^{p} N_i u_i$$

in the original equation, thus

$$I = \int_{\Omega_e} \left[\Im\left(\sum_{i=1}^{p} N_i u_i \right) - g \right]^2 d\Omega_e = \int_{\Omega_e} \left[\frac{d^2 \sum_{1}^{p} N_i u_i}{dx^2} + \alpha \frac{d \sum_{1}^{p} N_i u_i}{dx} - g \right]^2 dx \tag{2.104}$$

and (b) minimization of the derived functional as

$$\frac{\partial I}{\partial u_i} = 2 \int_{\Omega_e} \left[\Im\left(\sum_{i=1}^{p} N_i u_i \right) - g \right] \left(\frac{d^2 N_i}{dx^2} + \alpha \frac{dN_i}{dx} \right) dx = 0, \quad i = 1, m \tag{2.105}$$

where m is the total number of unknowns in the problem. Equation (2.105) can be regarded as a weighted residual statement in which the weighting function is in terms of the derivatives of shape functions in a form that reflects the original differential equation. Combination of this equation with the weighted residual statement derived from the standard Galerkin method (Equation (2.48)), yields a Petrov–Galerkin formulation for the original equation. Further explanations about this point are given in Chapter 4.

2.2.5   Solution of time-dependent problems

Important classes of polymeric flow processes are described by time-dependent differential equations. The most convenient method for solution of the time-dependent differential equations by finite element procedures is the 'partial discretization' technique. In this technique, the space–time domain is not discretized as a whole and instead, time derivatives are treated separately. In addition, finite element discretization of temporal derivatives is usually avoided and instead more direct methods are used. The θ time-stepping technique and the Taylor–Galerkin method are the most frequently used partial discretization procedures for the solution of transient problems by weighted residual finite element schemes.

The θ time-stepping method

In this technique, initially, the time derivatives in a differential equation are kept unchanged and the spatial discretization is carried out to form a weighted residual statement in the usual manner. Therefore after the spatial discretization, instead of a set of algebraic equations which are normally derived for steady-state problems, a system of ordinary differential equations in terms of time derivatives are generated. In general, for the class of single step θ methods this system is shown using matrix notation as

$$[M]_\theta\{\dot{X}\}_\theta + [K]_\theta\{X\}_\theta = \{F\}_\theta \tag{2.106}$$

where the subscript θ indicates that the weighted residuals statement is derived at time level θ $(0 \leq \theta \leq 1)$, as is shown in Figure 2.29.

The temporal derivative term in Equation (2.106) is approximated by a forward difference as

$$\{\dot{X}\}_\theta = \frac{\{X\}_{n+\theta\Delta t} - \{X\}_n}{\theta\Delta t} = \frac{\{X\}_{n+1} - \{X\}_n}{\Delta t} \tag{2.107}$$

The remaining terms in Equation (2.106) are approximated using a linear interpolation as

$$\{A\}_\theta = (1 - \theta)\{A\}_n + \theta\{A\}_{n+1} \tag{2.108}$$

Therefore

$$[K]_\theta\{X\}_\theta = (1 - \theta)[K]_n\{X\}_n + \theta[K]_{n+1}\{X\}_{n+1} \tag{2.109}$$

$$\{F\}_\theta = (1 - \theta)\{F\}_n + \theta\{F\}_{n+1} \tag{2.110}$$

Substitution from Equations (2.107), (2.108) and (2.109–2.110) into Equation (2.106) and carrying out algebraic manipulations gives

$$([M]_\theta + \theta\Delta t[K]_{n+1})\{X\}_{n+1} = ([M]_\theta - (1 - \theta)\Delta t[K]_n)\{X\}_n +$$
$$((1 - \theta)\{F\}_n + \theta\{F\}_{n+1})\Delta t \tag{2.111}$$

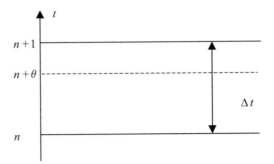

Figure 2.29   Time stepping scheme

The algebraic system given as Equation (2.111) represents the working equation of the θ method. On the basis of this equation a global set is derived and solved to obtain the unknowns at time level $n + 1$ using the known values at time level n.

The described method can generate a first-order backward or a first-order forward difference scheme depending whether $\theta = 0$ or $\theta = 1$ is used. For $\theta = 0.5$, the method yields a second order accurate central difference scheme, however, other considerations such as the stability of numerical calculations should be taken into account. Stability analysis for this class of time stepping methods can only be carried out for simple cases where the coefficient matrix in Equation (2.106) is symmetric and positive-definite (i.e. self-adjoint problems; Zienkiewicz and Taylor, 1994). Obviously, this will not be the case in most types of engineering flow problems. In practice, therefore, selection of appropriate values of θ and time increment Δt is usually based on trial and error. Factors such as the nature of non-linearity of physical parameters and the type of elements used in the spatial discretization usually influence the selection of the values of θ and Δt in a problem.

Equation (2.106) gives rise to an implicit scheme except for $\theta = 0$. The application of implicit schemes for transient problems yields a set of simultaneous equations for the field unknown at the new time level $n + 1$. As can be seen from Equation (2.111) some of the terms in the coefficient matrix should also be evaluated at the new time level. Therefore application of the described scheme requires the use of iterative algorithms. Various techniques for enhancing the speed of convergence in these algorithms can be found in the literature (Pittman, 1989).

Taylor–Galerkin method

Consider a partial differential equation, representing a time dependent flow problem given as

$$\frac{\partial u(\boldsymbol{x}, t)}{\partial t} + \Im[u(\boldsymbol{x}, t)] - g = 0 \tag{2.112}$$

where $\Im$ is a linear differential operator with respect to the spatial variables $\boldsymbol{x}$. Taylor series expansion of the field unknown, $u(\boldsymbol{x},t)$, in Equation (2.112) with respect to the time increment, gives

$$u^{n+1} = u^n + \Delta t \frac{\partial u}{\partial t}\Big|^n + \frac{1}{2}(\Delta t)^2 \frac{\partial^2 u}{\partial t^2}\Big|^n + \dots \tag{2.113}$$

Time derivatives in expansion (2.113) can now be substituted using the differential equation (2.112) (Donea, 1984). The first order time derivative in expansion (2.113) is substituted using Equation (2.112) as

$$\frac{\partial u(\boldsymbol{x}, t)}{\partial t} = -\Im[u(\boldsymbol{x}, t)] + g \tag{2.114}$$

Repeated differentiation of Equation (2.114) with respect to the time variable also gives the higher-order time derivatives of the unknown, for example

$$\frac{\partial^2 u(\boldsymbol{x}, t)}{\partial t^2} = \frac{\partial}{\partial t}\{-\Im[u(\boldsymbol{x}, t)] + g\} \tag{2.115}$$

Any first-order temporal derivative of $u(\boldsymbol{x},t)$ appearing on the right-hand side of Equation (2.115) can again be substituted from Equation (2.114). All of the temporal derivatives in expansion (2.113) can in principle be found using the original differential equation. This provides a differential equation that is exclusively in terms of the spatial variables. This equation can be discretized and solved by the Galerkin method in the usual manner. In practice, Taylor series expansion of the field variable is truncated and usually only the first few terms are kept. Accuracy of the time-stepping scheme in the Taylor–Galerkin method is therefore dependent on the highest order of the time derivative remaining in the expansion after its truncation. For example, a second-order scheme should involve the second-order time derivative of the field unknown. However, repeated differentiation and substitution of the temporal derivatives using the governing differential equations of complex field problems may prove to be difficult. A computationally more efficient form of the second-order Taylor–Galerkin scheme based on a time-split procedure has been developed by Townsend and Webster (1987) which can resolve this difficulty.

REFERENCES

Bathe, K. J., 1996. *Finite Element Procedures*, Prentice Hall, Englewood Cliffs, NJ.

Brenner, S. C. and Scott, L. R., 1994. *The Mathematical Theory of Finite Element Methods*, Springer-Verlag, New York.

Bogner, F. K., Fox, F. L. and Schmit, L. A., 1965. The generation of interelement-compatible stiffness and mass matrices by the use of interpolation formulae. *Proc. Conf. on Matrix Methods in Structural Mechanics*, Air Force Institute of Technology, Wright-Patterson AF Base, OH.

Brooks, A. N. and Hughes, T. J. R., 1982. Streamline-upwind/Petrov Galerkin formulations for convection dominated flows with particular emphasis on the incompressible Navier–Stokes equations. *Comput. Methods Appl. Mech. Eng.* **32**, 199–259.

Ciarlet, P. G., 1978. *The Finite Element Method for Elliptic Problems*, North-Holland, Amsterdam.

Crouzeix, M. and Raviart, P. A., 1973. Conforming and non-conforming finite elements for solving the stationary Navier–Stokes equations. *RAIRO, Serie Rouge* **3**, 33–76.

Donea, J., 1984. A Taylor–Galerkin method for convective transport problems. *Int. J. Num. Methods Eng.* **20**, 101–119.

Forsythe, G. E. and Meler, C. B., 1967. *Computer Solution of Linear Algebraic Systems*, Prentice Hall, Englewood Cliffs, NJ.

Gerald, C. F. and Wheatley, P. O., 1984. *Applied Numerical Analysis*, 3rd edn. Addison-Wesley, Reading, MA.

Girault, V. and Raviart, P. A., 1986. *Finite Element Methods for Navier–Stokes Equations*, Springer-Verlag, Berlin.

Hinton, E. and Owen, D. R. J., 1977. *Finite Element Programming*, Academic Press, London.

Hood, P., 1976. Frontal solution program for unsymmetric matrices. *Int. J. Numer. Methods Eng.* **10**, 379–399.

Hughes, T. J. R. and Brooks, A. N., 1979. A multidimensional upwind scheme with no cross-wind diffusion. In: Hughes, T. J. R. (ed.), *Finite Element Methods for Convection Dominated Flows*, AMD Vol. 34, ASME, New York.

Irons, B. and Ahmad, S., 1980. *Techniques of Finite Elements*, ch. 13, Ellis Horwood/Wiley, Chichester, pp. 215–244.

Johnson, C., 1987. *Numerical Solution of Partial Differential Equations by the Finite Element Method*, Cambridge University Press, Cambridge.

Lapidus, L. and Pinder, G. F., 1982. *Numerical Solution of Partial Differential Equations in Science and Engineering*, Wiley, New York.

Mitchell, A. R. and Wait, R., 1977. *The Finite Element Method in Partial Differential Equations*, Wiley, London.

Petera, J., Nassehi, V. and Pittman, J. F. T., 1989. Petrov–Galerkin methods on iso-parametric bilinear and biquadratic elements tested for a scalar convection–diffusion problem. *Int. J. Numer. Meth. Heat Fluid Flow* **3**, 205–222.

Pironneau, O., 1989. Finite element methods for fluids. Wiley, Chichester.

Pittman, J. F. T., 1989. Finite elements for field problems. In: Tucker III, C. L. (ed.), *Computer Modeling for Polymer Processing*, ch. 6, Hanser Publishers, Munich, pp. 237–331.

Pittman, J. F. T. and Nakazawa, S., 1984. Finite element analysis of polymer processing operations. In: Pittman, J. F. T., Zienkiewicz, O. C., Wood, R. D. and Alexander, J. M. (eds), *Numerical Analysis of Forming Processes*, Wiley, Chichester.

Reddy, J. N., 1993. *An Introduction to the Finite Element Method*, 2nd edn, McGraw-Hill, New York.

Strang, G. and Fix, G. J., 1973. *An Analysis of the Finite Element Method*, Prentice Hall, Englewood Cliffs, NJ.

Taylor, C. and Hood, P., 1973. A numerical solution of the Navier–Stokes equations using the finite element technique. *Comput. Fluids* **1**, 73–100.

Townsend, P. and Webster, M. F., 1987. An algorithm for the three dimensional transient simulation of non-Newtonian fluid flow. In: Pande, G. N. and Middleton, J. (eds), *Transient Dynamic Analysis and Constitutive Laws for Engineering Materials Vol. 2*, T12, Nijhoff-Holland, Swansea, pp. 1–11.

Zienkiewicz, O. C. and Morgan, K., 1983. *Finite Elements and Approximation*, Wiley, New York.

Zienkiewicz, O. C. and Taylor, R. L., 1994. *The Finite Element Method*, 4th edn, Vols 1 and 2, McGraw-Hill, London.

3

Finite Element Modelling of Polymeric Flow Processes

Weighted residual finite element methods described in Chapter 2 provide effective solution schemes for incompressible flow problems. The main characteristics of these schemes and their application to polymer flow models are described in the present chapter.

As already discussed, in general, polymer flow models consist of the equations of continuity, motion, constitutive and energy. The constitutive equation in generalized Newtonian models is incorporated into the equation of motion and only in the modelling of viscoelastic flows is a separate scheme for its solution required.

Equations of continuity and motion in a flow model are intrinsically connected and their solution should be described simultaneously. Solution of the energy and viscoelastic constitutive equations can be considered independently.

3.1 SOLUTION OF THE EQUATIONS OF CONTINUITY AND MOTION

Application of the weighted residual method to the solution of incompressible non-Newtonian equations of continuity and motion can be based on a variety of different schemes. In what follows general outlines and the formulation of the working equations of these schemes are explained. In these formulations Cauchy's equation of motion, which includes the extra stress derivatives (Equation (1.4)), is used to preserve the generality of the derivations. However, velocity and pressure are the only field unknowns which are obtainable from the solution of the equations of continuity and motion. The extra stress in Cauchy's equation of motion is either substituted in terms of velocity gradients or calculated via a viscoelastic constitutive equation in a separate step.

The convection term in the equation of motion is kept for completeness of the derivations. In the majority of low Reynolds number polymer flow models this term can be neglected.

3.1.1 The U–V–P scheme

U–V–P schemes belong to the general category of mixed finite element techniques (Zienkiewicz and Taylor, 1994). In these techniques both velocity and pressure in the governing equations of incompressible flow are regarded as primitive variables and are discretized as unknowns. The method is named after its most commonly used two-dimensional Cartesian version in which U, V and P represent velocity components and pressure, respectively. To describe this scheme we consider the governing equations of incompressible non-Newtonian flow (Equations (1.1) and (1.4), Chapter 1) expressed as

$$\begin{cases} \nabla . v = 0 \\ \rho \dfrac{\partial v}{\partial t} + \rho v . \nabla v = -\nabla p \delta + \nabla . \tau + \rho g \end{cases} \tag{3.1}$$

where v, ρ, τ and g represent velocity, fluid density, pressure, extra stress and body force, respectively, and δ is the Kronecker delta. In the U–V–P method, weighted residual statements of the above equations over element Ω_e, in the discretized domain are formulated as

$$\begin{cases} \displaystyle\int_{\Omega_e} M_L \nabla . \bar{v} \, d\Omega_e = 0 \\ \displaystyle\int_{\Omega_e} \rho N_J \left(\dfrac{\partial \bar{v}}{\partial t} + \bar{v}^0 \nabla \bar{v} \right) d\Omega_e = \int_{\Omega_e} N_J (-\nabla \bar{p} \delta + \nabla . \bar{\tau} + \rho g) d\Omega_e \end{cases} \tag{3.2}$$

where an over bar indicates the approximated (i.e. elementally interpolated) variables, M_L and N_J are appropriate weight functions, $\bar{v}^0$ is assumed to represent the velocity found at a previous iteration step to linearise the convection term in the discretized equation of motion. Equation set (3.2) is based on the general equation (2.46) representing a weighted residual finite element statement. The most immediate requirement in the application of the U–V–P scheme to the modelling of incompressible flow regimes is the satisfaction of a stability condition known as the Babuska–Brezzi or BB criterion (Babuska, 1971; Brezzi, 1974). Using a simple (and mathematically non-rigorous) explanation it can be said that this requirement arises from the absence of a pressure term in the incompressible continuity equation. In the application of the finite element technique to incompressible flow it was found that the U–V–P method in conjunction with elements generating identical interpolations for velocity and pressure yields inaccurate and oscillatory results. These oscillations were shown to disappear when elements belonging to the Taylor–Hood or Crouzeix–Raviart family, that provide unequal order interpolations for velocities and pressure, are used (Zienkiewicz et al., 1986). Therefore the discretized velocity and pressure shown in Equation (3.2) are expressed as

$$\begin{cases} \bar{\mathbf{v}} = \sum_{I=1}^{n} N_I(\mathbf{x}) v_I \\[2em] \bar{p} = \sum_{L=1}^{m} M_L(\mathbf{x}) p_L \end{cases} \tag{3.3}$$

Further details of the BB, sometimes referred to as Ladyzhenskaya–Babuska–Brezi (LBB) condition and its importance in the numerical solution of incompressible flow equations can be found in textbooks dealing with the theoretical aspects of the finite element method (e.g. see Reddy, 1986). In practice, the instability (or checker-boarding) of pressure in the U–V–P method can be avoided using a variety of strategies.

The main strategies for obtaining stable results by the U–V–P scheme for incompressible flow are as follows:

- To use non-standard elements belonging to the Taylor–Hood or Crouzeix–Raviart groups that satisfy the BB condition. Examples of useful elements in this category are given in Table 3.1, for further explanations about the properties of these elements see Pittman (1989).

- To use an element that, although does not satisfy the BB condition, is capable of filtering out parasitic oscillations (Lee *et al.*, 1979).

Table 3.1

Element	Interpolation		Number of nodes and order of continuity	
	Velocity	Pressure	Velocity	Pressure
Triangular Taylor–Hood	Quadratic	Linear	6 Vertices and mid-sides C^0	3 Vertices C^0
Rectangular Taylor–Hood	Bi-quadratic	Bi-linear	9 Corners, mid-sides and centre C^0	4 Corners C^0
Triangular Crouzeix–Raviart	Quadratic	Constant	6 Vertices and mid-sides C^0	1 Centre C^{-1}
Rectangular Crouzeix–Raviart	Bi-quadratic	Linear	9 Corners, mid-sides and centre C^0	1 Centre C^{-1}

- To use a technique that can circumvent the necessity for satisfaction of the BB condition.

Algorithms based on the last approach usually provide more flexible schemes than the other two methods and hence are briefly discussed in here. Hughes *et al.* (1986) and de Sampaio (1991) developed Petrov–Galerkin schemes based on equal order interpolations of field variables that used specially modified weight functions to generate stable finite element computations in incompressible flow. These schemes are shown to be the special cases of the method described in the following section developed by Zienkiewicz and Wu (1991).

3.1.2   The U–V–P scheme based on the slightly compressible continuity equation

As already explained the necessity to satisfy the BB stability condition restricts the types of available elements in the modelling of incompressible flow problems by the U–V–P method. To eliminate this restriction the continuity equation representing the incompressible flow is replaced by an equation corresponding to slightly compressible fluids, given as

$$\frac{1}{\rho c^2}\frac{\partial p}{\partial t} + \nabla.\boldsymbol{v} = 0 \tag{3.4}$$

where c is the speed of sound in the fluid. In this case the discretized form of the governing flow equations are formulated as

$$\begin{cases} \displaystyle\int_{\Omega_e} N_J\left(\frac{1}{\rho c^2}\frac{\partial \bar{p}}{\partial t} + \nabla.\bar{\boldsymbol{v}}\right)\mathrm{d}\Omega_e = 0 \\[2em] \displaystyle\int_{\Omega_e} \rho N_J\left(\frac{\partial \bar{\boldsymbol{v}}}{\partial t} + \bar{\boldsymbol{v}}^0\nabla\bar{\boldsymbol{v}}\right)\mathrm{d}\Omega_e = \int_{\Omega_e} N_J(-\nabla\bar{p}\boldsymbol{\delta} + \nabla.\bar{\boldsymbol{\tau}} + \rho\boldsymbol{g})\mathrm{d}\Omega_e \end{cases} \tag{3.5}$$

Using different types of time-stepping techniques Zienkiewicz and Wu (1991) showed that equation set (3.5) generates naturally stable schemes for incompressible flows. This resolves the problem of mixed interpolation in the U–V–P formulations and schemes that utilise equal order shape functions for pressure and velocity components can be developed. Steady-state solutions are also obtainable from this scheme using iteration cycles. This may, however, increase computational cost of the solutions in comparison to direct simulation of steady-state problems.

3.1.3 Penalty schemes

The penalty method is based on the expression of pressure in terms of the incompressibility condition (i.e. the continuity equation) as

$$p = -\lambda(\nabla.\boldsymbol{v}) \tag{3.6}$$

where λ is a very large number called the penalty parameter. Equation (3.6), which represents a perturbed form of the continuity equation, is used to substitute the pressure in the equation of motion in terms of velocity gradients. This can, in physical terms, be interpreted as easing of the incompressibility constraint to consider instead, a slightly compressible flow regime. Depending whether the described substitution of pressure is carried out before or after the discretization of the governing equations, two different types of the penalty method are developed. Regardless of which formulation is used the elimination of pressure as a prime unknown yields a more compact set of working equations and hence the total computational cost of the penalty methods is less than comparable U–V–P schemes. The main drawback of the penalty method is the generation of ill-conditioned equations that result from the multiplication of some of the terms in the stiffness matrix by a large number. Iterative methods designed to cope with ill-conditioned equations may be needed to improve the performance of the penalty schemes (Zienkiewicz *et al.*, 1985). In general, in the application of the penalty schemes to the polymer flow problems the following points should be considered:

- Level of enforcement of the incompressibility condition depends on the magnitude of the penalty parameter. If this parameter is chosen to be excessively large then the working equations of the scheme will be dominated by the incompressibility constraint and may become singular. On the other hand, if the selected penalty parameter is too small then the mass conservation will not be assured. In non-Newtonian flow problems, where shear-dependent viscosity varies locally, to enforce the continuity at the right level it is necessary to maintain a balance between the viscosity and the penalty parameter. To achieve this the penalty parameter should be related to the viscosity as $\lambda = \lambda_0 \eta$ (Nakazawa *et al.*, 1982) where λ_0 is a large dimensionless parameter and η is the local viscosity. The recommended value for λ_0 in typical polymer flow problems is about 10^8.

- By using a variable penalty parameter related to local element size, round-off error in the solution of ill-conditioned finite element equations obtained in the penalty schemes can be reduced (Kheshgi and Scriven, 1985).

- Elimination of the pressure term from the equation of motion does not automatically yield a robust scheme for incompressible flow and it is still necessary to satisfy the BB stability condition by a suitable technique in both forms of the penalty method.

The continuous penalty technique

In the continuous penalty technique prior to the discretization of the governing equations, the pressure in the equation of motion is substituted from Equation (3.6) to obtain

$$\rho \frac{\partial \pmb{v}}{\partial t} + \rho \pmb{v}.\nabla \pmb{v} = -\nabla(-\lambda \nabla.\pmb{v})\pmb{\delta} + \nabla.\pmb{\tau} + \rho \pmb{g} \tag{3.7}$$

The discretization of Equation (3.7) gives

$$\int_{\Omega_e} \rho N_J \left(\frac{\partial \bar{\pmb{v}}}{\partial t} + \bar{\pmb{v}}^0 \nabla \bar{\pmb{v}}\right) d\Omega_e = \int_{\Omega_e} N_J \nabla(\lambda \nabla.\bar{\pmb{v}})\pmb{\delta} \, d\Omega_e + \int_{\Omega_e} N_J \nabla.\bar{\pmb{\tau}} \, d\Omega_e +$$

$$\int_{\Omega_e} \rho N_J \pmb{g} \, d\Omega_e \tag{3.8}$$

Equation (3.8) is the basic working equation of the continuous penalty method.
    As already explained, in order to enforce the incompressibility constraint some of the terms in the stiffness equation in the penalty method are multiplied by a very large parameter. In general, this yields an equation which is overwhelmed by its penalty terms and can only generate a trivial solution. To solve this problem, the penalty sub-matrix in the elemental coefficient matrix is forced to become singular. In practice, the penalty terms in the elemental coefficient matrix are calculated using a 'reduced integration' to achieve the required singularity (Zienkiewicz and Taylor, 1994). For example, in a two-dimensional problem all of the terms in the elemental stiffness matrix can be calculated using a 3 × 3 Gauss–Legendre quadrature except for the penalty terms that are found by a 2 × 2 procedure. This is said to be equivalent to using a lower order interpolation for pressure than the one used for velocity.

The discrete penalty technique

The use of selectively reduced integration to obtain accurate non-trivial solutions for incompressible flow problems by the continuous penalty method is not robust and failure may occur. An alternative method called the discrete penalty technique was therefore developed. In this technique separate discretizations for the equation of motion and the penalty relationship (3.6) are first obtained and then the pressure in the equation of motion is substituted using these discretized forms. Finite elements used in conjunction with the discrete penalty scheme must provide appropriate interpolation orders for velocity and pressure to satisfy the BB condition. This is in contrast to the continuous penalty method in which the satisfaction of the stability condition is achieved indirectly through the reduced integration of the penalty terms. The discrete penalty method combines the advantages of computational economy with the robustness of a numerical scheme in which the BB condition is directly satisfied.

3.1.4   Calculation of pressure in the penalty schemes – variational recovery method

The well-known inaccuracy of numerical differentiation precludes the direct calculation of pressure by the insertion of the computed velocity field into Equation (3.6). This problem is, however, very effectively resolved using the following 'variational recovery' method: Consider the discretized form of Equation (3.6) given as

$$\int_{\Omega_e} w\bar{p} \, d\Omega_e = -\int_{\Omega_e} w\lambda \nabla.\bar{v} \, d\Omega_e \tag{3.9}$$

where w is an appropriate weight function and the over bar means that the unknowns are approximated using the finite element shape functions in the usual manner. Equation (3.9) represents a variational statement corresponding to the penalty relation and its utilization in conjunction with the Galerkin finite element discretization yields a scheme for the calculation of nodal pressures. This is shown as

$$\int_{\Omega_e} N_I \sum N_K p_K \, d\Omega_e = -\lambda_0 \int_{\Omega_e} N_I \eta \nabla.\bar{v} \, d\Omega_e \tag{3.10}$$

where the penalty parameter in Equation (3.10) is substituted using Equation (3.6). It is important to note that the integrals on the right-hand sides of the equations arising from Equation (3.10) should be found at the reduced integration points. The coefficient matrix on the left-hand side of Equation (3.10) is the 'mass matrix' given as

$$M_{IK} = \int_{\Omega_e} N_I N_K \, d\Omega_e \tag{3.11}$$

This matrix is usually diagonalized using a simple mass lumping technique (Pittman and Nakazawa, 1984) to minimize the computational cost of pressure calculations in this method.

3.1.5   Application of Green's theorem – weak formulations

Normally, the extra stress in the equation of motion is substituted in terms of velocity gradients and hence this equation includes second order derivatives of velocity. Substitution of the pressure term via the penalty relation also gives rise to second-order derivatives of velocity. Therefore it appears that only finite elements whose shape functions guarantee inter-element continuity up to the first-order derivatives, at least, can be used to discretize the equation of motion. As explained in Chapter 2 the restriction on the order of continuity of permissible finite elements is readily relaxed (i.e. continuity requirement is weakened) by the application of Green's theorem to the second order derivatives. The general form of this application in the context of the described finite element schemes is as follows: Consider an integral as

$$I = \int\int_{\Omega} w\nabla.(\nabla f)d\Omega \tag{3.12}$$

Using integration by parts

$$\int\int_{\Omega} w\nabla.(\nabla f)d\Omega = \int\int_{\Omega} \nabla.(w\nabla f)d\Omega - \int\int_{\Omega} (\nabla w).(\nabla f)d\Omega \tag{3.13}$$

According to Green's theorem (Aris, 1989) we have

$$\int\int_{\Omega} \nabla.(w\nabla f)d\Omega = \int_{\Gamma} w\nabla f\bar{n}\ d\Gamma \tag{3.14}$$

where Γ is the boundary surrounding the domain Ω and $\bar{n}$ is the unit vector normal to Γ in the outward direction. Therefore

$$I = \int_{\Gamma} w\nabla f\bar{n}d\Gamma - \int\int_{\Omega} (\nabla w).(\nabla f)d\Omega \tag{3.15}$$

Therefore the second-order derivative of f appearing in the original form of I is replaced by a term involving first-order derivatives of w and f plus a boundary term. The boundary terms are, normally, cancelled out through the assembly of the elemental stiffness equations over the common nodes on the shared interior element sides and only appear on the outside boundaries of the solution domain. However, as is shown later in this chapter, the appropriate treatment of these integrals along the outside boundaries of the flow domain depends on the prescribed boundary conditions.

In practice, in order to maintain the symmetry of elemental coefficient matrices, some of the first order derivatives in the discretized equations may also be integrated by parts.

The described application of Green's theorem which results in the derivation of the 'weak' statements is an essential step in the formulation of robust U–V–P and penalty schemes for non-Newtonian flow problems.

3.1.6   Least-squares scheme

The basic procedure for the derivation of a least squares finite element scheme is described in Chapter 2, Section 2.4. Using this procedure the working equations of the least-squares finite element scheme for an incompressible flow are derived as follows:

Field unknowns in the governing flow equations are substituted using finite element approximations in the usual manner to form a set of residual statements. These statements are used to formulate a functional as

$$ F[(\bar{v}),\bar{p}] = \int\limits_{\Omega_e} \left\{ \left[\rho\frac{\partial\bar{v}}{\partial t} + \rho\bar{v}^0\nabla\bar{v} + \nabla\bar{p}\delta - \nabla.\bar{\tau} - \rho g \right]^2 + k[\nabla.\bar{v}]^2 \right\} d\Omega_e \qquad (3.16) $$

where an over bar represents approximation over an element Ω_e. The constant k is used to make the functional dimensionally consistent. Minimization of functional (3.16), with respect to field variables (i.e. velocity components etc.) leads to generation of the working equations of the least-squares scheme. As can be ascertained from functional (3.16) the working equations of this scheme will inevitably include second-order derivatives. Therefore only finite elements that can generate a sufficiently high-order discretization to cope with second-order derivatives can be used in conjunction with this scheme. This is a severe set back for this method that can restrict its general applicability in flow modelling. However, the least-squares method has been advocated as a powerful technique that provides better numerical stability than either U–V–P or penalty schemes in a wide range of fluid dynamical problems (Bell and Surana, 1994).

3.2   MODELLING OF VISCOELASTIC FLOW

In a significant number of polymer processes the influence of fluid elasticity on the flow behaviour is small and hence it is reasonable to use the generalized Newtonian approach to analyse the flow regime. In generalized Newtonian fluids the extra stress is explicitly expressed in terms of velocity gradients and viscosity and can be eliminated from the equation of motion. This results in the derivation of Navier–Stokes equations with velocity and pressure as the only prime field unknowns. Solution of Navier–Stokes equations (or Stokes equation for creeping flow) by the finite element schemes is the basis of computer modelling of non-elastic polymer flow regimes. In contrast, in viscoelastic flow models the extra stress can only be given through implicit relationships with the rate of strain, and hence remains as a prime field unknown in the governing equations. In this case therefore, in conjunction with the governing equations of continuity and momentum (generally given as Cauchy's equation of motion) an appropriate constitutive equation must be solved. Numerical solution of viscoelastic constitutive equations has been the subject of a considerable amount of research in the last two decades. This has given rise to a plethora of methods that apply to various types of differential or integral constitutive equations. Despite significant achievements of the last two decades a universal modelling methodology, that can generate stable and accurate results for the range of Weissenberg numbers observed in practical flow processes, has not been developed. It is generally accepted that the complex non-linear nature of visco-elastic fluids precludes the development of a universal method for all situations and efforts should be focused on solving individual problems.

The main categories of numerical algorithms used to solve viscoelastic flow problems are summarized by Keunings (1989). These schemes are primarily divided into 'coupled' and 'decoupled' techniques. In coupled methods, the governing equations of continuity, motion and rheology for a viscoelastic flow regime are solved simultaneously to obtain the whole set of velocity, pressure and stress variables as the prime field unknowns. These techniques usually use an iterative procedure such as Newton's method. In contrast, in the decoupled methods the equations of continuity and motion are solved separately from the rheological equation. The main procedure in these techniques is to start from a 'known' flow kinematics to calculate the viscoelastic extra stress by solving the rheological equation. The calculated stress field is then inserted into the equation of motion and the velocity and pressure fields are found. This process is iterated until a converged solution is obtained.

The coupled techniques are more readily applicable to the differential visco-elastic constitutive equations and their extension to the integral models is not common. The main advantage of the coupled methods is that, potentially, they can benefit from the quadratic rate of convergence of Newton's iteration scheme. Decoupled methods, on the other hand, make more efficient use of the available computer core capacity. However, the rate of convergence of the decoupled methods is generally slow. The important point to note is that both techniques require considerable computer CPU times if solutions that reflect realistic conditions are attempted. Numerous examples of both categories of coupled and decoupled methods in conjunction with various types of finite element formulations and multiplicity of viscoelastic constitutive models can be found in the literature. However, it has not been possible to establish the accuracy, or clearly demonstrate the advantageous, of some of the elaborate schemes that have been reported. A thorough description or critical evaluation of various classes of viscoelastic flow models is beyond the scope of the present book and the interested reader should refer to scientific papers and books published about this topic.

In general, the utilization of 'integral models' requires more elaborate algor-ithms than the differential viscoelastic equations. Furthermore, models based on the differential constitutive equations can be more readily applied under 'general' conditions.

Difficulties involved in the solution of viscoelastic constitutive equations have so far prevented the development of a modelling methodology with general applicability for these regimes. Using specially modified Petrov–Galerkin, tech-niques (Hughes et al., 1986; Hughes, 1987), numerically stable results for these equations can be generated. However, it is not possible to demonstrate the experimental accuracy of many of these solutions. Further research is still required to develop more efficient and reliable finite element schemes for the modelling of viscoelastic flow regimes under the conditions that arise in areas such as polymer processing.

In the following section representative examples of the development of finite element schemes for most commonly used differential and integral viscoelastic models are described.

3.2.1   Outline of a decoupled scheme for the differential constitutive models

The first finite element schemes for differential viscoelastic models that yielded numerically stable results for non-zero Weissenberg numbers appeared less than two decades ago. These schemes were later improved and shown that for some benchmark viscoelastic problems, such as flow through a two-dimensional section with an abrupt contraction (usually a width reduction of four to one), they can generate simulations that were 'qualitatively comparable' with the experimental evidence. A notable example was the coupled scheme developed by Marchal and Crochet (1987) for the solution of Maxwell and Oldroyd constitutive equations. To achieve stability they used element subdivision for the stress approximations and applied inconsistent streamline upwinding to the stress terms in the discretized equations. In another attempt, Luo and Tanner (1989) developed a typical decoupled scheme that started with the solution of the constitutive equation for a fixed-flow field (e.g. obtained by initially assuming non-elastic fluid behaviour). The extra stress found at this step was subsequently inserted into the equation of motion as a pseudo-body force and the flow field was updated. These authors also used inconsistent streamline upwinding to maintain the stability of the scheme.

In this section the discretization of upper-convected Maxwell and Oldroyd-B models by a modified version of the Luo and Tanner scheme is outlined. This scheme uses the subdivision of elements suggested by Marchal and Crochet (1987) to generate smooth stress fields (Swarbrick and Nassehi, 1992a).

In the absence of body force, the dimensionless form of the governing model equations for two-dimensional steady-state incompressible creeping flow of a viscoelastic fluid are written as

$$\begin{cases} \nabla \cdot \boldsymbol{\sigma} = 0 \\ \nabla \cdot \boldsymbol{v} = 0 \end{cases} \tag{3.17}$$

In Equation (3.17), the Cauchy stress is defined as

$$\boldsymbol{\sigma} = -p\boldsymbol{\delta} + \boldsymbol{\tau}_{\mathrm{v}} + \boldsymbol{\tau}_{\mathrm{e}} \tag{3.18}$$

where p is pressure, δ is the Kronecker delta, τ_v and τ_e are, respectively, the viscous and elastic parts of the extra stress and v is velocity. The viscous part of the extra stress is given in terms of the rate of deformation as $\tau_v = r(\nabla v + \nabla v^T)$, where r is a dimensionless material parameter defined as

$$r = \eta_s/(\eta_s + \eta_m) \tag{3.19}$$

where η_m is the viscosity of the viscoelastic fluid and η_s is the viscosity of the 'Newtonian solvent' (see Chapter 1).

The elastic part of the extra stress for upper-convected Maxwell (UCM) and Oldroyd-B fluids is written as (see Chapter 1)

$$\tau_e + Ws\frac{\Delta_{-100}\tau_e}{\Delta t} = (1-r)(\nabla v + \nabla v^T) \tag{3.20}$$

where Ws is the Weissenberg number. The governing equations for the upper-convected Maxwell flow are obtained by putting $r = 0$, while the Oldroyd-B case is represented by $r = 0.11$ (Crochet, 1982).

After the substitution of Cauchy stress via Equation (3.20) and the viscous part of the extra stress in terms of rate of deformation, the equation of motion is written as

$$\nabla p\delta - r\nabla \cdot (\nabla v + \nabla v^T) = \nabla \cdot \tau_e \tag{3.21}$$

For simplicity, we define $T = \tau_e$ and $\overset{\triangledown}{T}$ ($\Delta_{-100}\tau_e/\Delta t$). As explained by Luo and Tanner (1989), the decoupled method requires a suitable variable transformation in the governing equations (3.20) and (3.21). This is to ensure that the discrete momentum equations always contain the real viscous term required to recover the Newtonian velocity–pressure formulation when Ws approaches zero. This is achieved by decomposing the extra stress T as

$$T = S + R \tag{3.22}$$

where

$$R = (1-r)(\nabla u + \nabla u^T). \tag{3.23}$$

Therefore Equations (3.20) and (3.21) are written in terms of S and R as

$$S + Ws\overset{\triangledown}{S} = -Ws\overset{\triangledown}{R} \tag{3.24}$$

and

$$\nabla p\delta - \nabla \cdot (\nabla u + \nabla u^T) = \nabla \cdot S \tag{3.25}$$

Derivation of the working equations

The weighted residual statement of Equation (3.25) over an element domain Ω_e is

$$\int_{\Omega_e} [\nabla p \boldsymbol{\delta} - \nabla \cdot (\nabla \boldsymbol{u} + \nabla \boldsymbol{u}^{\mathsf{T}})] N \, d\Omega_e = \int_{\Omega_e} (\nabla \cdot \boldsymbol{S}) N \, d\Omega_e \qquad (3.26)$$

where N is a weighting function. The integrand on the left-hand side of Equation (3.26) involves second-order derivatives of velocity and to preserve the inter-element continuity the velocity field needs to be approximated by appropriately high order interpolation functions. To weaken this requirement it is necessary to apply Green's theorem on Equation (3.26). This gives

$$\int_{\Omega_e} [(\nabla p \boldsymbol{\delta}) N + (\nabla \boldsymbol{u} + \nabla \boldsymbol{u}^{\mathsf{T}})] \cdot \nabla N] \, d\Omega_e = \int_{\Omega_e} (\nabla \cdot \boldsymbol{S}) N \, d\Omega_e +$$

$$\int_{\Gamma_e} (\nabla \boldsymbol{u} + \nabla \boldsymbol{u}^{\mathsf{T}}).\boldsymbol{n} N \, d\Gamma_e \qquad (3.27)$$

where $\boldsymbol{n}$ is the unit outward normal to the boundary Γ_e.

The weighted residual statements of Equations (3.17) and (3.25) are also formulated as

$$\int_{\Omega_e} (\nabla \cdot \boldsymbol{u}) N \, d\Omega_e = 0 \qquad (3.28)$$

and

$$\int_{\Omega_e} (\boldsymbol{S} + Ws \overset{\triangledown}{\boldsymbol{S}}) N \, d\Omega_e = - \int_{\Omega_e} Ws \overset{\triangledown}{\boldsymbol{R}} N \, d\Omega_e \qquad (3.29)$$

Depending on the type of elements used appropriate interpolation functions are used to obtain the elemental discretizations of the unknown variables. In the present derivation a mixed formulation consisting of nine-node bi-quadratic shape functions for velocity and the corresponding bi-linear interpolation for the pressure is adopted. To approximate stresses a 3×3 subdivision of the velocity–pressure element is considered and within these sub-elements the stresses are interpolated using bi-linear shape functions. This arrangement is shown in Figure 3.1.

The approximate forms of u, p and T are hence given as

$$\bar{\boldsymbol{u}} = \sum_I \boldsymbol{u}_I N_I, \quad \bar{p} = \sum_K p_K M_K, \quad \bar{\boldsymbol{T}} = \sum_L \boldsymbol{T}_L M_L \qquad (3.30)$$

where the letters N and M represent bi-quadratic and bi-linear shape functions, respectively. The subscripts I, K and L represent the nodes involved in the described approximations. After the substitution of the field unknowns in terms of their corresponding shape functions, from Equation (3.30), in the weighted residual statements of the governing equations and application of the Galerkin weighting the working equations of the scheme (i.e. the elemental stiffness equation) are formulated.

The momentum and continuity equations give rise to a 22 × 22 elemental stiffness matrix as is shown by Equation (3.31). In Equation (3.31) the subscripts I and J represent the nodes in the bi-quadratic element for velocity and K and L the four corner nodes of the corresponding bi-linear interpolation for the pressure. The weight functions, N_I and M_L, are bi-quadratic and bi-linear, respectively. The jth component of velocity at node J is shown as U_J^j. Summation convention on repeated indices is assumed. The discretization of the continuity and momentum equations is hence based on the U–V–P scheme in conjunction with a Taylor–Hood element to satisfy the BB condition.

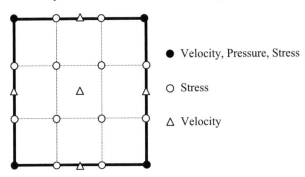

● Velocity, Pressure, Stress

○ Stress

△ Velocity

Figure 3.1 Subdivided element for the discretization of viscoelastic extra-stress

$$
\begin{bmatrix}
\int_{\Omega_e}(2N_{I,1}N_{J,1}+N_{I,2}N_{J,2})\mathrm{d}\Omega_e & \int_{\Omega_e}N_{I,2}N_{J,1}\ \mathrm{d}\Omega_e & \int_{\Omega_e}N_I M_{K,1}\ \mathrm{d}\Omega_e \\[2mm]
\int_{\Omega_e}N_{I,1}N_{J,2}\ \mathrm{d}\Omega_e & \int_{\Omega_e}(N_{I,1}N_{J,1}+2N_{I,2}N_{J,2})\ \mathrm{d}\Omega_e & \int_{\Omega_e}N_I M_{K,2}\ \mathrm{d}\Omega_e \\[2mm]
\int_{\Omega_e}M_L N_{J,1}\ \mathrm{d}\Omega_e & \int_{\Omega_e}M_L N_{J,2}\ \mathrm{d}\Omega_e & 0
\end{bmatrix}
\begin{bmatrix}
U_J^1 \\[2mm] U_J^2 \\[2mm] P_k
\end{bmatrix}
=
$$

$$
\begin{bmatrix}
\int_{\Omega_e}S_{1j,j}N_I\ \mathrm{d}\Omega_e + \int_{\Gamma_e}d_{1j}n_j N_I\ \mathrm{d}\Gamma_e \\[2mm]
\int_{\Omega_e}S_{2j,j}N_I\ \mathrm{d}\Omega_e + \int_{\Gamma_e}d_{2j}n_j N_I\ \mathrm{d}\Gamma_e \\[2mm]
0
\end{bmatrix}
\tag{3.31}
$$

In the decoupled scheme the solution of the constitutive equation is obtained in a separate step from the flow equations. Therefore an iterative cycle is developed in which in each iterative loop the stress fields are computed after the velocity field. The viscous stress R (Equation (3.23)) is calculated by the 'variational recovery' procedure described in Section 1.4. The elastic stress S is then computed using the working equation obtained by application of the Galerkin method to Equation (3.29). The elemental stiffness equation representing the described working equation is shown as Equation (3.32).

$$
\begin{bmatrix}
\int M_1 M_J\,d\Omega_e \\
+Ws\int u^m M_I^* M_{J,m}\,d\Omega_e \\
-2Ws\int u^1_{,1} M_I M_J\,d\Omega_e
\end{bmatrix}
\quad 0 \quad
2Ws\int u^1_{,2} M_I M_J\,d\Omega_e
\ \cdots
$$

$$(3.32)$$

The integrals in Equation (3.32) are found using a quadrature over the element domain Ω_e. The viscoelastic constitutive equations used in the described model are hyperbolic equations and to obtain numerically stable solutions the convection terms in Equation (3.32) are weighted using streamline upwinding as (inconsistent upwinding)

$$ M_I^* = M_I + \frac{\phi h^e}{|u|} u^J \cdot \frac{\partial M_I}{\partial x_J} \tag{3.33} $$

where $|u|$ is the magnitude of u, ϕ is a scaling factor and h^e is a characteristic element size defined as

$$ h^e = \sqrt{(x_{,\xi}+x_{,\eta})^2 + (y_{,\xi}+y_{,\eta})^2} \tag{3.34} $$

where (ξ, η) are local coordinates. The basic iterative procedure for the described decoupled scheme is given in the following chart.

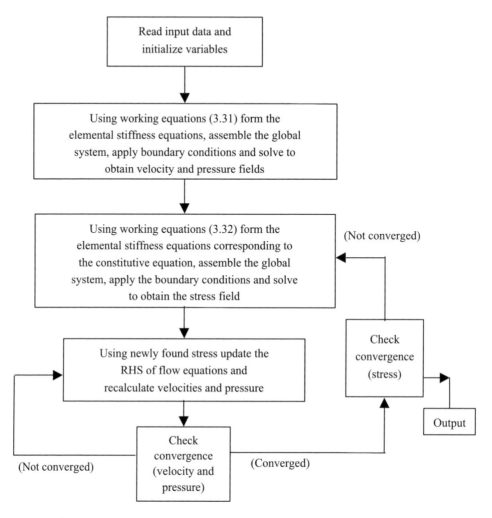

The outlined scheme is shown to yield stable solutions for non-zero Weissenberg number flows in a number of benchmark problems (Swarbric and Nassehi, 1992b). However, the extension of this scheme to more complex problems may involve modifications such as increasing of elemental subdivisions for stress calculations from 3 × 3 to 9 × 9 and/or the discretization of the stress field by bi-quadratic rather than bi-linear sub-elements. It should also be noted that satisfaction of the BB condition in viscoelastic flow simulations that use mixed formulations is not as clear as the case of purely viscous regimes.

3.2.2   Finite element schemes for the integral constitutive models

As mentioned in Chapter 1, in general, the solution of the integral viscoelastic models should be based on Lagrangian frameworks. In certain types of flow regimes, however, the solution of these equations can be based on an Eulerian approach. To give a brief description of the use of an Eulerian framework in the solution of integral constitutive models we consider the motion and straining of a fluid particle in a contracting domain as shown in Figure 3.2.

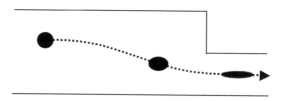

Figure 3.2 Deformation of a fluid particle along its trajectory in a contracting flow

If the position of the fluid particle at current time t is given by $x(t)$ then its motion is defined using the following position vector

$$x(t') = X\{[x(t), t], t'\} \tag{3.35}$$

where t' is a past time between $-\infty$ and t, hence the above vector gives the position of the fluid particle at historical time t'. The deformation gradient for this particle can now be defined as

$$F_t(t') = \frac{\partial X}{\partial x} \tag{3.36}$$

The right Cauchy–Green strain tensor corresponding to this deformation gradient is thus expressed as

$$C_t(t') = F_t^T(t') \cdot F_t(t') \tag{3.37}$$

The inverse of the Cauchy–Green tensor, C_t^{-1}, is called the Finger strain tensor. Physically the single-integral constitutive models define the viscoelastic extra stress τ_v for a fluid particle as a time integral of the deformation history, i.e.

$$\tau_v(t) = \int_{-\infty}^{t} m(t - t') S_t(t') \, dt' \tag{3.38}$$

where the deformation-dependent tensor S_t and its coefficient (the memory function) are generally defined by Equations (1.28) and (1.29), as described in Chapter 1. To develop an Eulerian solution scheme for the integral models, the deformation gradient defined by Equation (3.36), should be found explicitly in terms of the velocity components. To obtain such a relationship the following time derivative of the deformation gradient, describing the kinematics of the fluid particle

$$\frac{D}{Dt'}F_t(t') = \nabla v^T(t').F_t(t') \tag{3.39}$$

is integrated backward in time along the fluid particle path using the present state of the deformation as the initial condition. Although a general solution for Equation (3.39) is not available for certain classes of flow regimes it can be integrated to obtain the required explicit relationship between the Eulerian velocity field and the strain history. As an example, we consider the solution developed by Adachi (1983) for the class of two-dimensional steady-state flow systems whose streamlines can be represented by single-valued functions of only one of the spatial coordinates. Streamlines in a flow field represent the contour lines of the stream function ψ which is defined by the following relationships

$$\begin{cases} \dfrac{\partial \psi}{\partial x} = -v_y \\[2mm] \dfrac{\partial \psi}{\partial y} = v_x \end{cases} \tag{3.40}$$

where x and y are the spatial coordinates in a Cartesian coordinate system. We now define a Protean coordinate system as (Adachi, 1983)

$$\begin{cases} x_1 = x \\ x_2 = \psi(x, y) \end{cases} \tag{3.41}$$

In this system the coordinate x_2 along a streamline is constant. Therefore tracking of a fluid particle in this system is significantly simplified. This allows a closed-form solution of the basic kinematic equation (3.39) in the defined Protean coordinates. Using the coordinate transformation relationships given by Equation (3.41) the closed-form solution of the kinematical equation can be transformed into the Cartesian system in which the Eulerian velocity field is originally defined. For example the component of the Finger strain tensor is expressed as (Keunings, 1989)

$$C_{xx}^{-1}(t, t') = [v_x(t)^2 v_x(t')^2 + v_x(t)^2 v_y(t')^2]I^2 - \left[2\frac{v_x(t)^2 v_y(t')}{v_x(t')}\right]I + \frac{v_x(t)^2}{v_x(t')^2} \tag{3.42}$$

where I is an integral evaluated along the streamline as

$$I = \int_{t'}^{t} \frac{\partial v_x(\theta)}{\partial y} \frac{d\theta}{v_x^2(\theta)} \tag{3.43}$$

where the particle travel time along the streamline and the coordinate x are related on a one-to-one basis and

$$d\theta = \frac{dx(\theta)}{v_x(\theta)} \tag{3.44}$$

Therefore the Eulerian description of the Finger strain tensor, given in terms of the present and past position vectors x and x' of the fluid particle as $C_{xx}^{-1}(x, x')$, can now be expressed as

$$C_{xx}^{-1}(x, x') = [v_x(x)^2 v_x(x')^2 + v_x(x)^2 v_y(x')^2] I^2 - \left[2\frac{v_x(x)^2 v_y(x')}{v_x(x')}\right] I + \frac{v_x(x)^2}{v_x(x')^2}$$

$$\tag{3.45}$$

The integral I corresponding to Equation (3.45) is defined as

$$I = \int_{x'}^{x} \frac{\partial v_x(\xi)}{\partial y} \frac{d\xi}{v_x^3(\xi)} \tag{3.46}$$

which is evaluated along the streamline passing through the current position. The Eulerian form of the generic single-integral constitutive Equation (3.38) can now be defined as

$$\tau_v(x) = \int_{-\infty}^{x} m\left(\int_{x'}^{x} \frac{d\xi}{v_x(\xi)}\right) S(x, x') \frac{dx'}{v_x(x')} \tag{3.47}$$

Therefore the viscoelastic extra stress acting on a fluid particle is found via an integral in terms of velocities and velocity gradients evaluated upstream along the streamline passing through its current position. This expression is used by Papanastasiou et al. (1987) to develop a finite element scheme for viscoelastic flow modelling.

As mentioned earlier, the Eulerian solution of the single-integral viscoelastic constitutive equations can only be obtained for special classes of flow regimes and under general conditions a Lagrangian framework should be used. The Lagrangian systems developed for this purpose are similar to those used to track moving boundaries or free surfaces in a flow field. The development of a Lagrangian system for tracking fluid particle trajectories is discussed in a separate section later in this chapter.

3.2.3   Non-isothermal viscoelastic flow

The theoretical description of a non-isothermal viscoelastic flow presents a conceptual difficulty. To give a brief explanation of this problem we note that in a non-isothermal flow field the evolution of stresses will by affected be the temperature distribution. Let us now consider a viscoelastic fluid undergoing stress relaxation. The rate of stress relaxation is determined internally within the fluid, according to an internal timescale or 'clock'. From an independent observer's point of view, as the temperature rises the molecular motion within the fluid in a unit of time increases. Thus the internal timescale of the fluid has become shorter to allow a faster rate of relaxation. To evaluate the temperature dependency of material parameters a method called 'time–temperature shifting' is used that takes into account the variations of the internal timescale of the fluid. Through the application of this technique the fluid properties at a temperature T are found on the basis of a master curve given at a reference temperature of T_0. However, in a flow regime a fluid particle experiences various states of temperature and hence for the incorporation of time–temperature shifting into a material property model a systematic approach is needed. Such an approach has been developed for specific situations. For example, for linear viscoelastic boundary value problems a method called the Morland–Lee hypothesis is used (Tanner, 2000). According to this hypothesis, a pseudo-time based on the time measured by the internal clock of the fluid is defined. The relationship between the pseudo-time given by ξ and the observer's time t is expressed as

$$d\xi = \frac{dt}{a_T(T)} \tag{3.48}$$

where $a_T(T)$ is the time-shifting factor. Integration of Equation (3.48) gives

$$\xi = \int_0^t a_T^{-1}[T(t')]dt' \tag{3.49}$$

which defines the relationship between the particle's time and the time of the observation. The pseudo-time concept is utilized to obtain the non-isothermal forms of the viscoelastic constitutive equations. This kind of variable transformation is more convenient using the integral constitutive equations, however, differential non-isothermal models have also been developed.

3.3   SOLUTION OF THE ENERGY EQUATION

The thermal conductivity of polymeric fluids is very low and hence the main heat transport mechanism in polymer processing flows is convection (i.e. corresponds to very high Peclet numbers; the Peclet number is defined as $\rho c U l / k$ which represents the ratio of convective to conductive energy transport). As emphasized before, numerical simulation of convection-dominated transport phenomena by the standard Galerkin method in a fixed (i.e. Eulerian) framework gives unstable and oscillatory results and cannot be used.

Streamline upwinding is the most commonly used technique in the development of stable finite element schemes for the solution of the energy equation in polymer flow models. In particular, consistent 'streamline upwind Petrov–Galerkin' schemes (see Chapter 2, Section 2.3) have gained widespread applications in many practical problems (Ghoreishy and Nassehi, 1997). It is also possible to use a moving (i.e. Lagrangian) framework to eliminate the convection term from the governing energy equation and generate stable solutions for heat transport in polymer processes without upwinding. However, application of Lagrangian schemes in practical problems should be based on algorithms such as adaptive regeneration of the finite element mesh (Morton, 1996) to prevent unacceptable element distortions and in general is more difficult than the use of upwinding.

Derivation of the working equations of upwinded schemes for heat transport in a polymeric flow is similar to the previously described weighted residual Petrov–Galerkin finite element method. In this section a basic outline of this derivation is given using a steady-state heat balance equation as an example.

Assuming constant physical coefficients for simplicity, the steady-state energy equation is expressed as

$$\rho c v_i \frac{\partial T}{\partial x_i} - k \delta_{ij} \frac{\partial^2 T}{\partial x_i^2} - \dot{q} = 0 \tag{3.50}$$

where summation convention over the repeated index i is used. In Equation (3.50), ρ, c and k are constant density, specific heat and conductivity, respectively, v_i represents the components of the velocity field, T is temperature, $\dot{q} = \eta \dot{\gamma}^2$ is viscous heat dissipation rate and δ_{ij} is the Kronecker delta.

Starting with the usual discretization step the field unknowns in Equation (3.50) are replaced by their approximate interpolated representation (e.g. $\bar{T} = \sum_{I=1}^{p} N_I T_I$) over elemental domain (Ω_e). The formulated residual statement is weighted and integrated to give

$$\int_{\Omega_e} \rho c W_J \bar{v}_i \frac{\partial \bar{T}}{\partial x_i} \, d\Omega_e - \int_{\Omega_e} W_{JJ} k \delta_{ij} \frac{\partial^2 \bar{T}}{\partial x_i^2} \, d\Omega_e - \int_{\Omega_e} W_{JJ} \dot{q} \, d\Omega_e = 0 \tag{3.51}$$

where W_J and W_{JJ} are weight functions. After the application of Green's theorem to its second-term Equation (3.51) yields

$$\int_{\Omega_e} \rho c W_J \bar{v}_i \frac{\partial \bar{T}}{\partial x_i} \, d\Omega_e + \int_{\Omega_e} k \delta_{ij} \frac{\partial W_{JJ}}{\partial x_i} \frac{\partial \bar{T}}{\partial x_i} \, d\Omega_e - \int_{\Omega_e} W_{JJ} \dot{q} \, d\Omega_e = \int_{\Gamma_e} k n_i W_{JJ} \frac{\partial \bar{T}}{\partial x_i} \, d\Gamma_e \tag{3.52}$$

where Γ_e is the elemental domain boundary and n_i is the component of outward unit vector normal to the boundary line. The weight functions in the inconsistent streamline upwind method are given as

$$W_J = N_J + \alpha N_{J,i} \frac{\bar{v}_i}{|v|^2} \quad \text{(see Equation (2.79)}$$ (3.53)

and

$$W_{JJ} = N_J$$ (3.54)

where $N_J (J = 1, \ldots, p)$ represents elemental shape functions.

In the consistent streamline upwind Petrov–Galerkin (SUPG) scheme all of the terms in Equation (3.52) are weighted using the function defined by Equation (3.53) and hence $W_{JJ} = W_J$.

It is evident that in the consistent upwinding method the weight function applied to the second-order (conduction) term in the energy equation involves derivatives of the shape functions. Therefore in this case the weight function corresponding to the conduction term in the weak form of the discretized energy equation includes second-order derivatives of shape functions. Consequently, in the consistent SUPG schemes the Laplacian of the shape functions at elemental integration points should be calculated. This is not a trivial calculation and, in particular, in discretizations that use isoparametric mapping requires considerable algebraic manipulations (Petera *et al.*, 1993).

In multi-dimensional problems the upwinding parameter, α (Equation (3.53)), cannot be defined through a rigorous relationship. Using heuristic extensions of the one-dimensional optimal streamline upwinding formula, a number of equations for multi-dimensional domains have been derived (Zienkiewicz and Taylor, 1994). All of these relationships inevitably involve coefficients that do not have clear physical definitions. In practice, therefore, the selection of the best upwinding parameter for a multi-dimensional heat transport problem in polymer processing should be based on numerical experiments involving trial and error procedures. It is recommended to maintain the level of upwinding just above the threshold of instability.

Extension of the streamline Petrov–Galerkin method to transient heat transport problems by a space–time least-squares procedure is reported by Nguen and Reynen (1984). The close relationship between SUPG and the least-squares finite element discretizations is discussed in Chapter 4. An analogous transient upwinding scheme, based on the previously described θ time-stepping technique, can also be developed (Zienkiewicz and Taylor, 1994).

In general, thermal properties of polymers are temperature dependent and may change significantly during a flow process. Therefore in order to maintain the accuracy of the finite element solution of the energy equation, it may be necessary to include a procedure in the scheme that allows the specific heat c, and conductivity k, to vary with temperature. In practical solution algorithms thermal coefficients in the energy equation are repeatedly updated, according to the latest computed temperature field, and the solution is iterated. However, updating of these parameters by a simple iterative loop in which the entire process involving evaluation of elemental stiffness equations, assembly and solution of the global system is repeated in each cycle is not computationally efficient. In addition to the large number of calculations involved in such a cycle, the rate of convergence of the iterations can also be very low. To achieve computing efficiency the required iteration loop should hence be modified. The product approximation method developed by Christie *et al.* (1981) provides an efficient algorithm for the reduction of computational time in the finite element solution of non-linear problems. Using this method, thermal coefficients in the energy equation are discretized in conjunction with temperature. Therefore in the global set they appear as nodal values multiplied by a constant and can be easily updated according to the corresponding nodal temperatures. Other techniques have also been developed to avoid repeated evaluation and assembly of the right-hand side in the solution of the energy equation (e.g. see Cardona and Idelsohn, 1986; Nour-Omid, 1987).

3.4 IMPOSITION OF BOUNDARY CONDITIONS IN POLYMER PROCESSING MODELS

Identification and prescription of appropriate boundary conditions is a crucial step in the development of finite element models for engineering problems. In numerical simulation of polymer flow processes, these conditions may include velocity, stress (surface force components), temperature and a datum for pressure. It is clear that in the case of viscoelastic models, where the flow inside a domain is influenced by the history of fluid deformation before it enters through an inlet, all ambiguity about the boundary conditions cannot be resolved. Therefore, in practical (engineering) simulations of viscoelastic flow a set of conditions that can be shown to make the best possible physical sense, under the given conditions, is usually used. Complete mathematical evaluations of such boundary conditions are, in general, not possible. Generalized Newtonian behaviour is often considered as a useful limiting case that provides an indication for the general validity of the imposed boundary conditions in polymer processing models. In this section common types of boundary conditions used in these models are described.

3.4.1 Velocity and surface force (stress) components

Consider the weighted residual statement of the equation of motion in a steady state Stokes flow model, expressed as

$$\int_{\Omega} N_J (\nabla \cdot \bar{\boldsymbol{\sigma}} - \rho \boldsymbol{g}) \mathrm{d}\Omega = 0 \qquad (3.55)$$

where N_J is a weight function, $\boldsymbol{\sigma}$ is the Cauchy stress (the over bar represents approximation within Ω) and $\rho\boldsymbol{g}$ is the body force. Application of Green's theorem to Equation (3.55) gives

$$\int_{\Omega} \nabla N_J \cdot \bar{\boldsymbol{\sigma}}\ \mathrm{d}\Omega - \int_{\Gamma} N_J \bar{\boldsymbol{\sigma}} \cdot \boldsymbol{n}\ \mathrm{d}\Omega - \int_{\Omega} N_J \rho\boldsymbol{g}\ \mathrm{d}\Omega = 0 \qquad (3.56)$$

where Γ is the boundary surrounding Ω and $\boldsymbol{n}$ is the unit vector normal to Γ in the outward direction. Substituting for $\boldsymbol{\sigma}$ in the first term of Equation (3.56) via Equation (1.3) gives

$$\int_{\Omega} \nabla N_J \cdot \bar{\boldsymbol{\tau}}\ \mathrm{d}\Omega - \int_{\Omega} \nabla N_J \cdot \bar{p}\boldsymbol{\delta}\ \mathrm{d}\Omega - \int_{\Gamma} N_J \bar{\boldsymbol{\sigma}} \cdot \boldsymbol{n}\ \mathrm{d}\Omega - \int_{\Omega} N_J \rho\boldsymbol{g}\ \mathrm{d}\Omega = 0 \qquad (3.57)$$

where $\boldsymbol{\tau}$ is the extra stress, p is pressure and $\boldsymbol{\delta}$ is the Kronecker delta. In a generalized Newtonian flow the extra stress is given in terms of the rate of deformation tensor as $\boldsymbol{\tau} = \eta(\nabla\boldsymbol{v} + \nabla\boldsymbol{v}^{\mathrm{T}})$, where η is fluid viscosity. Therefore the weighted residual statement of the described flow model consisting of the equations of continuity and motion is written as

$$\begin{cases} \displaystyle\int_{\Omega} M_L (\nabla \cdot \bar{\boldsymbol{v}})\ \mathrm{d}\Omega = 0 \\[2ex] \displaystyle\int_{\Omega} \eta \nabla N_J \cdot (\nabla\bar{\boldsymbol{v}} + \nabla\bar{\boldsymbol{v}}^{\mathrm{T}})\ \mathrm{d}\Omega - \int_{\Omega} \nabla N_J \cdot \bar{p}\boldsymbol{\delta}\ \mathrm{d}\Omega - \int_{\Omega} N_J \cdot \rho\boldsymbol{g}\ \mathrm{d}\Omega - \int_{\Gamma} N_J \bar{\boldsymbol{\sigma}} \cdot \boldsymbol{n}\ \mathrm{d}\Omega = 0 \end{cases}$$

$$(3.58)$$

where M_L is a weight function. Equation (3.58) provides the basis for the formulation of elemental equations in the previously described finite element schemes. Imposition of boundary conditions in incompressible flow models is now described using this equation without any significant loss of generality. Methods that may be used to prescribe the, usually unknown, boundary conditions at the entry to a domain in viscoelastic models are also briefly discussed. Consider the two-dimensional domain shown in Figure 3.3 where $\Gamma = \Gamma_1 \cup \Gamma_2 \cup \Gamma_3 \cup \Gamma_4$.

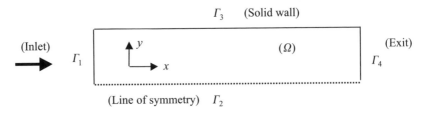

Figure 3.3   Boundary lines in a flow domain

Inlet conditions

Typically velocity components along the inlet are given as essential (also called Dirichlet)-type boundary conditions. For example, for a flow entering the domain shown in Figure 3.3 they can be given as

$$\begin{cases} v_y = 0 \\ v_x = v_x(y) \end{cases}$$

on Γ_1. To impose this type of boundary condition the specified values are inserted into the equations where they appear and rows corresponding to the given degrees of freedom are eliminated from the global set of finite element equations.

In general it is beneficial to use finer mesh divisions near a solid wall.

In viscoelastic models in addition to the described conditions, stresses at the inlet should be given. As already mentioned there is no universal method to define such conditions, however, the following options may be considered (Tanner, 2000):

- Adding time-dependent terms to the equations the simulation is treated as an initial value problem in which at a given reference time all stresses are zero; the steady-state solution can be found iteratively.

- Adding sections before the actual domain inlet arbitrary conditions are imposed at the artificial starting point and then iterated to obtain required stresses.

- Using a known solution at the inlet. To provide an example for this option, let us consider the finite element scheme described in Section 2.1. Assuming a 'fully developed' flow at the inlet to the domain shown in Figure 3.3, v_y, $(\partial v_y / \partial y) = 0$ and by the incompressibility condition $(\partial v_x / \partial x) = 0$, x derivatives of all stress components are also zero. Therefore at the inlet the components of the equation of motion (3.25) are reduced to

$$\begin{cases} \dfrac{\partial p}{\partial x} - \dfrac{\partial^2 v_x}{\partial y^2} = \dfrac{\partial S_{xy}}{\partial y} \\ \dfrac{\partial p}{\partial y} = \dfrac{\partial S_{yy}}{\partial y} \end{cases}$$

hence the constitutive equation (3.24) is reduced to

$$\begin{cases} S_{xx} - 2Ws\dfrac{\partial v_x}{\partial y}S_{xy} = 2Ws\dfrac{\partial v_x}{\partial y}R_{xy} \\ S_{yy} = 0 \\ S_{xy} = Ws\dfrac{\partial v_x}{\partial y}R_{yy} \end{cases}$$

Also Equation (3.23) yields

$$\begin{cases} R_{xx} = 0 \\ R_{yy} = 0 \\ R_{xy} = (1 - r)\dfrac{\partial v_x}{\partial y} \end{cases}$$

at the inlet. Therefore $S_{xy} = 0$ and $S_{xx} = 2Ws(1 - r)\,(\partial v_x/\partial y)^2$, where the x component of velocity is usually given as a parabolic function in terms of y.

Line of symmetry

The normal component of velocity and tangential component of surface force are set to zero along a line of symmetry. For the domain shown in Figure 3.3 these are expressed as

$$\begin{cases} v_y = 0 \\ \sigma_x = 0 \end{cases}$$

on Γ_2. Imposition of the first condition is identical to the procedure used for prescribing inlet velocity components and the second condition is simply satisfied by setting the boundary line integral in the discretized equation of motion (i.e. $\int_\Gamma N_J \cdot \bar{\sigma}.n \, d\Gamma$ in Equation (3.58) to zero. Note that in this example components of n, unit vector normal to the line of symmetry are $(0,-1)$ and consequently the term containing σ_{xx} in the boundary line integral is multiplied by zero and eliminated. The remaining term, i.e. the shearing force component, is $-\sigma_{xy} = -\tau_{xy} = -\eta[(\partial v_x/\partial y) + (\partial v_y/\partial x)]$ which is reduced to $-\eta(\partial v_x/\partial y)$ along Γ_2. Therefore imposition of $\sigma_x = 0$ is equivalent to prescribing a natural (von Neumann)-type boundary condition for the axial component of the velocity.

In general, lines of symmetry within a flow field may not be parallel to the coordinate axes. Boundary conditions along these lines will still be physically similar to the conditions used in the simple example given here, but their imposition will not be trivial. In this case boundary integrals corresponding to these lines should be evaluated and additional algebraic relationships representing zero normal velocity conditions (i.e. $v.n = 0$ across a line of symmetry) should be incorporated into the global set of equations.

Solid walls

On no-slip walls zero velocity components can be readily imposed as the required boundary conditions ($v_x = v_y = 0$ on Γ_3 in the domain shown in Figure 3.3). Details of the imposition of slip-wall boundary conditions are explained later in Section 4.2.

Exit conditions

Typically the exit velocity in a flow domain is unknown and hence the prescription of Dirichlet-type boundary conditions at the outlet is not possible. However, at the outlet of sufficiently long domains fully developed flow conditions may be imposed. In the example considered here these can be written as

$$\begin{cases} v_y = 0 \\ \sigma_x = 0 \end{cases}$$

Imposition of the first condition is as discussed before and the second condition is again satisfied by setting the boundary line integral (i.e. $\int_\Gamma N_J \cdot \bar{\sigma}.n \, d\Gamma$ in Equation (3.58) to zero. We note that along the exit line components n of are $(1,0)$ and hence in this case the normal component of the surface force, expressed as $\sigma_{xx} = -p + 2\eta(\partial v_x/\partial x)$ will be set to zero. In addition, along the exit line $(\partial v_y/\partial y) = 0$, which using the incompressible continuity equation gives $(\partial v_x/\partial x) = 0$. Therefore setting $\sigma_{xx} = 0$ implies a pressure datum of zero at the domain outlet. This also shows that if the normal component of surface force is not given at any part of the domain boundary then zero pressure at a single node may be imposed as a datum. The setting of pressure boundary conditions is, in general, inconsistent with the incompressibility constraint and should be avoided. In the specific case where the pressure in a single node is prescribed, the continuity equation relating to that node should be removed from the set of discretized equations to avoid algebraic inconsistency. The validity of imposing fully developed exit boundary conditions in a flow model depends on factors such as the type of elements used, number of element layers between the inlet and outlet, inlet and wall boundary conditions and generally on fluid viscosity. Inappropriate imposition of developed flow conditions at a domain outlet reduces the accuracy of the solution and may give rise to spurious oscillations (Gresho *et al.*, 1980). In the flow domains that are not considered to be long enough to impose developed flow conditions, stress-free conditions at the domain outlet may be used. In this case, both shear and normal components of the surface forces at the exit are set to zero. This is again satisfied by setting of the boundary integral along the exit line to zero.

To avoid imposition of unrealistic exit boundary conditions in flow models Taylor *et al.* (1985) developed a method called 'traction boundary conditions'. In this method starting from an initial guess, outflow condition is updated in an iterative procedure which ensures its consistency with the flow regime immediately upstream. This method is successfully applied to solve a number of turbulent flow problems.

Papanastasiou *et al.* (1992) suggested that in order to generate realistic solutions for Navier–Stokes equations the exit conditions should be kept 'free' (i.e. no outflow conditions should be imposed). In this approach application of Green's theorem to the equations corresponding to the exit boundary nodes is avoided. This is equivalent to imposing 'no exit conditions' if elements with shape functions of the order at least equal to that of the differential equation are used. In discretizations using lower-order elements a boundary condition based on setting highest order differentials to zero is imposed (Sani and Gresho, 1994). Renardy (1997) has shown that the 'free' or 'no outlet' method provides a well-defined problem in the context of discretized governing equations.

3.4.2   Slip-wall boundary conditions

Imposition of no-slip velocity conditions at solid walls is based on the assumption that the shear stress at these surfaces always remains below a critical value to allow a complete wetting of the wall by the fluid. This implies that the fluid is constantly sticking to the wall and is moving with a velocity exactly equal to the wall velocity. It is well known that in polymer flow processes the shear stress at the domain walls frequently surpasses the critical threshold and fluid slippage at the solid surfaces occurs. Wall-slip phenomenon is described by Navier's slip condition, which is a relationship between the tangential component of the momentum flux at the wall and the local slip velocity (Silliman and Scriven, 1980). In a two-dimensional domain this relationship is expressed as

$$[\beta\boldsymbol{\tau}.\boldsymbol{n} + (\boldsymbol{v} - \boldsymbol{v}_b)].\boldsymbol{t}^\mathrm{T} = 0 \tag{3.59}$$

where $\boldsymbol{t}$ and $\boldsymbol{n}$ are unit vectors tangent and normal to the boundary, $\boldsymbol{\tau}$ is the extra stress tensor, β is a slip coefficient, $\boldsymbol{v}$ is the fluid velocity vector and $\boldsymbol{v}_b$ is the velocity of the solid wall. Equation (3.59) together with the following equation which represents no flow through a solid wall, are used to impose slip-wall boundary conditions.

$$\boldsymbol{v}.\boldsymbol{n} = 0 \tag{3.60}$$

Consider a solid wall section as is shown in Figure 3.4. The following relationships between the components of unit outward normal and tangential vectors are true at all points

$$t_y = +n_x \qquad t_x = -n_y \tag{3.61}$$

Equations (3.59) and (3.60) are recast in terms of their components and solved together. After algebraic manipulations and making use of relations (3.61) slip-wall velocity components are found as

$$u - u_b = -\eta\beta n_y\left[2\left(\frac{\partial u}{\partial x} - \frac{\partial v}{\partial y}\right)n_x n_y + \left(\frac{\partial u}{\partial y} + \frac{\partial v}{\partial x}\right)(n_y^2 - n_x^2)\right] \tag{3.62}$$

$$v - v_b = -\eta\beta n_x\left[2\left(\frac{\partial u}{\partial x} - \frac{\partial v}{\partial y}\right)n_x n_y + \left(\frac{\partial u}{\partial y} + \frac{\partial v}{\partial x}\right)(n_y^2 - n_x^2)\right] \tag{3.63}$$

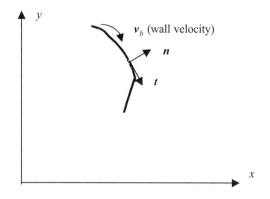

Figure 3.4   Slip at a solid wall

The slip coefficient β is defined as

$$\beta = \beta_0 / l \tag{3.64}$$

where β_0 is the initial slip coefficient and l is a characteristic flow domain dimension. The limit of $\beta \to 0$ corresponds to no-slip ($u = u_b$, $v = v_b$) and the limit of $\beta \to \infty$ gives the perfect slip condition. In general, the coefficient β depends on the invariants of the stress tensor and surface roughness. Navier's slip condition can be discretized in a manner similar to the main flow equations and directly incorporated into the finite element working equations (Ghoreishy and Nassehi, 1997) to impose a slip-wall condition.

3.4.3   Temperature and thermal stresses (temperature gradients)

Normally, in the finite element solution of the energy equation, temperatures at the inlet and solid walls of a flow domain and zero thermal stresses along the lines of symmetry and exit are specified as the boundary conditions. Imposition of these conditions is straightforward and is carried out according to the procedures explained for the prescription of similar types of boundary conditions in the solution of flow equations. In problems where the temperature at a downstream boundary is specified, i.e. a 'hard' exit boundary condition is to be imposed, preferential mesh refinement near the outlet is often necessary to maintain accuracy. This type of 'hardness' may also be eased if the problem is recast in a different frame that provides a simpler formulation for the model equations. For example, a cylindrical (r,θ) coordinate system provides a more natural framework than a Cartesian coordinate system to model the heat transfer associated with a tangential viscous flow in an annulus. Easing of the boundary conditions by using more appropriate frameworks can lead to more accurate solutions or better computing economy.

In the finite element solution of the energy equation it is sometimes necessary to impose heat transfer across a section of the domain wall as a boundary condition in the process model. This type of convection (Robins) boundary condition is given as

$$q_b = h(T - T_a) \qquad (3.65)$$

where q_b is the boundary heat flux, h is the heat transfer coefficient, T is the boundary surface temperature and T_a is the outside temperature. In the discretized energy equation (e.g. see Equation (3.52)) boundary heat flux is represented by the boundary line integrals. Therefore at the boundaries representing heat transfer surfaces these integrals do not vanish and should be evaluated using Equation (3.65). In most instances it is simpler to avoid the evaluation of these integrals and impose the required condition by a technique based on the use of 'virtual' elements. In this technique a thin layer of elements is attached to the outside of heat transfer boundaries. These elements are not part of the physical domain and hence referred to as the virtual elements. Heat flux through these elements is assumed to be in the direction normal to the surface (see Figure 3.5). Temperature variation across the virtual elements is also assumed to correspond to an analytic steady-state profile. These assumptions depend on using very thin elements and setting of the virtual specific heat to zero (Pittman, 1989). Heat conduction through the layer of virtual elements can now be related to the real surface heat flux as

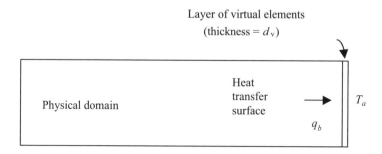

Layer of virtual elements
(thickness $= d_v$)

Heat transfer surface

Physical domain

T_a

q_b

Figure 3.5   Virtual element layer for the imposition of boundary heat flux

$$h(T - T_a) = \frac{k_v}{d_v}(T - T_a) \qquad (3.66)$$

where k_v and d_v are the conductivity and thickness of the virtual elements, respectively. The virtual conductivity is hence expressed as

$$k_v = d_v h \qquad (3.67)$$

On a curved surface the virtual conductivity is found analogously as

$$k_v = d_v \frac{R}{R_{lm}} h \tag{3.68}$$

where R and R_{lm} are the local radius of curvature of the boundary surface and the log-mean radius of the boundary and virtual surfaces, respectively. In practice, therefore, the outside temperature is set as an essential condition at the virtual element boundaries and the evaluation of line integrals at the real boundary is avoided.

Descriptions given in Section 4 of this chapter about the imposition of boundary conditions are mainly in the context of finite element models that use C^0 elements. In models that use Hermite elements derivatives of field variable should also be included in the set of required boundary conditions. In these problems it is necessary to ensure that appropriate 'normality' and 'tangentiality' conditions along the boundaries of the domain are satisfied (Petera and Pittman, 1994).

3.5   FREE SURFACE AND MOVING BOUNDARY PROBLEMS

In these problems, flow geometry is not known *a priori* and some sections of the domain boundary may change with flow. This situation arises in a variety of polymer processes such as injection moulding and mixing in partially filled chambers. Free surface flow regimes are also encountered in extrusion and wire coating operations where die swell is a common phenomenon. Various techniques for the modelling of free boundary flow regimes have been developed in the last two decades. Some of these methods are process specific, or they were developed in conjunction with particular numerical schemes and cannot be regarded as general simulation tools.

Modelling of steady-state free surface flow corresponds to the solution of a boundary value problem while moving boundary tracking is, in general, viewed as an initial value problem. Therefore, classification of existing methods on the basis of their suitability for boundary value or initial value problems has also been advocated.

The general class of free boundary flow problems can, however, be modelled using the volume of fluid (VOF) approach (Nichols *et al.*, 1980). The main concept in this technique is to solve, simultaneously with the governing flow equations, an additional equation that represents the unknown boundary. Three different versions of this method are described in the following sections.

3.5.1   VOF method in 'Eulerian' frameworks

In a fixed two-dimensional Cartesian coordinate system, the continuity equation for a free boundary is expressed as

$$\frac{\partial F}{\partial t} + v_x \frac{\partial F}{\partial x} + v_y \frac{\partial F}{\partial y} = 0 \tag{3.69}$$

where v_x and v_y are the flow velocity components and F is defined as a material density function. This function has a value of unity in filled sections of the flow domain and is zero outside of the free boundary. At the free boundary itself this function has a value between 0 and 1, (usually 0.4–0.5; Petera and Nassehi, 1996). The hyperbolic partial differential Equation (3.69) describes transient convection of a free boundary in a flow field. In models based on stationary (Eulerian) frameworks the solution of this equation usually requires streamline upwinding.

Straightforward application of the VOF method in domains that include irregular and curved boundaries is rather complicated. To resolve this problem a more flexible version of the original method has been developed by Thompson (1986). In this technique the free boundary flow regime is treated as a two-phase system. The phases are assumed to consist of the fluid filled and void regions, respectively. The free boundary is regarded as the interface separating these phases. To model the flow field in this manner, voids are assumed to contain a virtual fluid represented by a set of virtual physical properties. Hence, the technique is referred to as the 'pseudo-density' method. In practice, physical coefficients in the governing flow equations are expressed as

$$\chi = \chi_f F + \chi_v (1 - F) \tag{3.70}$$

where χ is a given physical parameter (e.g. density) and χ_f and χ_v are the values of this parameter in the fluid filled and void regions, respectively. The free surface density function is 1 in fluid filled sections and is 0 in the voids. Hence using Equation (3.70) appropriate coefficients are automatically inserted into the working equations and the solution for the entire domain is readily obtained. On the free boundary itself this equation yields the interpolated values of the physical coefficients. The solution scheme starts from a known distribution of the surface function values. At the end of each time step new values of F are found and the position of the interface between the phases (i.e. the free boundary) is updated. Despite its simplicity the effectiveness of the 'pseudo-density' method is restricted by the necessity to use artificial physical parameters in parts of the flow domain.

3.5.2   VOF method in 'Arbitrary Lagrangian–Eulerian' frameworks

The geometrical flexibility of the VOF scheme can be significantly improved if in its formulation, instead of using a fixed framework, a combination of a Lagrangian–Eulerian approach is adopted. The most common approach to develop such a combined framework is the application of the Arbitrary

Lagrangian–Eulerian (ALE) method. In the ALE technique the finite element mesh used in the simulation is moved, in each time step, according to a pre-determined pattern. In this procedure the element and node numbers and nodal connectivity remain constant but the shape and/or position of the elements change from one time step to the next. Therefore the solution mesh appears to move with a velocity which is different from the flow velocity. Components of the mesh velocity are time derivatives of nodal coordinate displacements expressed in a two-dimensional Cartesian system as

$$
\begin{cases}
\dfrac{\partial x}{\partial t} = v_{mx} \\[2mm]
\dfrac{\partial y}{\partial t} = v_{my}
\end{cases}
\tag{3.71}
$$

Governing flow equations, originally written in an Eulerian framework, should hence be modified to take into account the movement of the mesh. The time derivative of a variable f in a moving framework is found as

$$
\left.\frac{\partial f}{\partial t}\right|_{M} = \left.\frac{\partial f}{\partial x}\right|_{M} v_{mx} + \left.\frac{\partial f}{\partial y}\right|_{M} v_{my} + \left.\frac{\partial f}{\partial t}\right|_{E}
\tag{3.72}
$$

where indices M and E refer to moving and fixed frameworks, respectively. Substitution of the Eulerian time derivatives via relationships based on Equation (3.72) gives the modified form of the governing equations required in the ALE scheme. The free boundary equation is therefore expressed as

$$
\frac{\partial F}{\partial t} + (v_x - v_{mx})\frac{\partial F}{\partial x} + (v_y - v_{my})\frac{\partial F}{\partial y} = 0
\tag{3.73}
$$

The streamline upwinding method is usually employed to obtain the discretized form of Equation (3.73). The solution algorithm in the ALE technique is similar to the procedure used for a fixed VOF method. In this technique, however, the solution found at the end of the nth time step, based on mesh number n, is used as the initial condition in a new mesh (i.e. mesh number $n + 1$). In order to minimize the error introduced by this approximation the difference between the mesh configurations at successive computations should be as small as possible. Therefore the time increment should be small. In general, adaptive or re-meshing algorithms are employed to construct the required finite element mesh in successive steps of an ALE procedure (Donea, 1992). In some instances it is possible to generate the finite element mesh required in each step of the computation in advance, and store them in a file accessible to the computer program. This can significantly reduce the CPU time required for the simulation (Nassehi and Ghoreishy, 1998). An example in which this approach is used is given in Chapter 5.

3.5.3   VOF method in 'Lagrangian' frameworks

In a Lagrangian framework the coordinate system in which the governing flow equations are formulated moves with the flow field. Therefore flow equations written in such a system do not include any convection terms. In particular, the free surface continuity equation (Equation (3.69)) is reduced to a simple time derivative expressed as $(\partial F/\partial t) = 0$. However, the integration of this derivative is not trivial because in a free boundary domain it should be evaluated between variable (i.e. time-dependent) limits. In addition, a method should be adopted to prevent computational mesh distortions that will naturally occur in a Lagrangian framework. Various methods based on adaptive re-meshing are used to achieve this objective (e.g. see Morton, 1996). However, some of these techniques lack geometrical flexibility (e.g. they are only suitable for domains that do not include curved boundaries) or they require high CPU times. A robust and geometrically flexible method with implementation that requires relatively moderate CPU times is described in this section. This method is based on the tracking of fluid particle trajectories passing through each node in the computational domain and provides an effective technique for regeneration of the mesh at the end of each time step (Petera and Nassehi, 1996). The free surface continuity equation can be readily integrated using this approach.

Consider the position of a material point in a flow field described by the following position vector

$$X = f(p, t) \tag{3.74}$$

where p is the position at the current time t. At a given reference time, t^{r}, the particle occupies a specific position defined as

$$x = f(p, t^{r}) \tag{3.75}$$

Any given material point can only occupy a single position at a time and hence Equation (3.75) can be used to find the position of the point as

$$p = f^{-1}(x, t^{r}) \tag{3.76}$$

The substitution for p from Equation (3.76) into Equation (3.74) gives

$$X = f[f^{-1}(x, t^{r}), t] = \tilde{X}(x, t^{r}; t) \tag{3.77}$$

If the reference time t^{r} and the current time t coincide then the reference and current positions will also coincide and the right-hand side of Equation (3.77) can be replaced by the reference position defined as x in Equation (3.76). In a velocity field given as $u = u(x, t^{r})$ the motion of a material point can be described as

$$\frac{dX(x, t^{r}; t)}{dt} = u[X(x, t^{r}; t), t] \tag{3.78}$$

The distance covered by a fluid particle in this flow field in a time interval of $\Delta t_n = t_{n+1} - t_n$ can be found by integrating Equation (3.78) as

$$x - X^n = \int_{t_n}^{t_{n+1}} u[X(t), t] \, dt \tag{3.79}$$

Using the mean value theorem for definite integrals

$$x - X^n = u[X(t_\theta), t_\theta].\Delta t_n \tag{3.80}$$

where θ is a parameter between zero and one. Using Taylor series expansion the velocity function in the neighbourhood of the point (x, t_n) is given as

$$u[X(t_\theta), t_\theta] = u(x, t_n) + \partial_t u(x, t_n).(t_\theta - t_n) + \nabla u(x, t_n).[X(t_\theta) - x] \tag{3.81}$$

The second and third terms on the right-hand side of Equation (3.81) are approximated using known values at a backward time step as

$$\partial_t u(x, t_n) \cong \frac{u(x, t_n) - u(x, t_{n-1})}{\Delta t_{n-1}} \tag{3.82}$$

$$t_\theta - t_n \cong \tfrac{1}{2}\Delta t_n \tag{3.83}$$

$$X(t_\theta) - x \cong \tfrac{1}{2}[X(t_n) - x] \tag{3.84}$$

After the substitution from Equations (3.82), (3.83) and (3.84) into Equation (3.81) and in turn substituting from the resultant relationship into Equation (3.80) and rearranging the following equation describing the trajectory of fluid particles is found

$$x - X = \frac{\Delta t_n}{2} \left(\delta + \tfrac{1}{2}\Delta t_n \nabla u_n\right)^{-1} \left[u_n + \tfrac{1}{2}\frac{\Delta t_n}{\Delta t_{n-1}}(u_n - u_{n-1})\right] \tag{3.85}$$

where δ is the Kronecker delta. The solution algorithm based on Equation (3.85) consists of the following steps:

Step 1

A domain that can be safely assumed to represent the entire flow field is selected and discretized into a 'fixed' mesh of finite elements. The part of this domain that is filled by fluid is called the 'current' mesh. Nodes within the current mesh are regarded as 'active' nodes. At the active nodes located inside the current mesh the surface density function $F = 1$. This function for the nodes on the free boundary is normally given as 0.5.

Step 2

Location arrays showing the numbers of elements that contain each given node in the fixed mesh and its boundary are prepared and stored in a file.

Step 3

Appropriate initial conditions on all nodes are given.

Step 4

Time variable is incremented.

Step 5

Governing flow equations are solved with respect to the current domain.

Step 6

Using:

- coordinates of the nodes in the current domain (i.e. x)
- velocity field found at step 5 (i.e. u_n), and
- old time step velocity values (i.e. u_{n-1})

coordinates of the feet of the fluid particle trajectories (i.e. X) passing through nodal points are found via Equation (3.85).

Step 7

For each active node in the current mesh the corresponding location array is searched to find inside which element the foot of the trajectory currently passing through that node is located. This search is based on the solution of the following set of non-linear algebraic equations

$$X = \sum_I N_I(\xi, \eta) x_I$$

The unknowns in this equation are the local coordinates of the foot (i.e. ξ and η). After insertion of the global coordinates of the foot found at step 6 in the left-hand side, and the global coordinates of the nodal points in a given element in the right-hand side of this equation, it is solved using the Newton–Raphson method. If the foot is actually inside the selected element then for a quadrilateral element its local coordinates must be between -1 and $+1$ (a suitable criteria should be used in other types of elements). If the search is not successful then another element is selected and the procedure is repeated.

In the example shown in Figure 3.6, the trajectory passing through point A at the 'current time' is found to originate from the inside of element (e) at the previous time step.

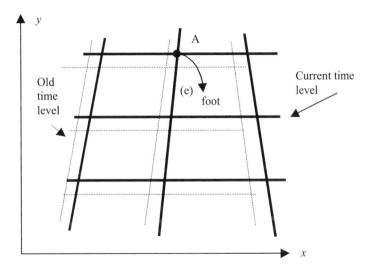

Figure 3.6   Determination of fluid particle trajectories

Step 8

After identification of the elements that contain feet of particle trajectories the old time step values of F at the feet are found by interpolating (or extrapolating for boundary nodes) its old time step nodal values. In the example shown in Figure 3.6 the old time value of F at the foot of the trajectory passing through A is found by interpolating its old nodal values within element (e).

Step 9

Free surface density functions calculated at step 8 are used as the initial conditions to update the current position of the surface using the following integration

$$I = \frac{d}{dt} \int_{\Omega_t} N_I F \, d\Omega \tag{3.86}$$

where Ω_t is a moving solution domain. Using Reynolds transport theorem (Aris, 1989) Equation (3.86) is expressed as

$$I = \frac{d}{dt} \int_{\Omega_t} N_I F \, d\Omega_t = \int_{\Omega_t} \left(\frac{DN_I}{Dt} F + N_I \frac{dF}{Dt} \right) d\Omega_t + \int_{\Omega_t} N_I F \nabla \cdot \boldsymbol{u} \, d\Omega_t = 0 \tag{3.87}$$

The right-hand side of Equation (3.87) is set to zero considering that DN_I/Dt, DF/Dt and the divergence of the velocity field in incompressible fluids are all equal to zero. Therefore, after integration Equation (3.87) yields

$$\int_{\Omega_{n+1}} (N_I F)^{n+1} \, d\Omega_{n+1} = \int_{\Omega_n} (N_I F)^n \, d\Omega_n \tag{3.88}$$

Steps 4 to 9 are repeated until the end of required simulation time.

An example describing the application of this algorithm to the finite element modelling of free surface flow of a Maxwell fluid is given in Chapter 5.

REFERENCES

Adachi, K., 1983. Calculation of strain histories in Protean coordinate systems. *Rheol. Acta* **22**, 326–335.

Aris, R., 1989. *Vectors, Tensors and the Basic Equations of Fluid Mechanics*, Dover Publications, New York.

Babuska, B., 1971. Error bounds for finite element method. *Numer. Methods* **16**, 322–333.

Bell, B. C. and Surana, K. S., 1994. p-version least squares finite element formulations for two-dimensional, incompressible, non-Newtonian isothermal and non-isothermal fluid flow. *Int. J. Numer. Methods Fluids* **18**, 127–162.

Brezzi, F., 1974. On the existence, uniqueness and approximation of saddle point problems arising with Lagrange multipliers. *RAIRO, Serie Rouge* **8R-2**, 129–151.

Cardona, A. and Idelsohn, S., 1986. Solution of non-linear thermal transient problems by a reduction method. *Int. J. Numer. Methods Eng.* **23**, 1023–1042.

Christie, I. *et al.*, 1981. Product approximation for non-linear problems in the finite element method. *IMA J. Numer. Anal.* **1**, 253–266.

Crochet, M. J., 1982. Numerical simulation of die-entry and die-exit flow of a viscoelastic fluid. In: *Numerical Methods in Forming Processes*, Pineridge Press, Swansea.

de Sampaio, P. A. B., 1991. A Petrov–Galerkin formulation for the incompressible Navier–Stokes equations using equal order interpolation for velocity and pressure. *Int. J. Numer. Methods Eng.* **31**, 1135–1149.

Donea, J., 1992. Arbitrary Lagrangian–Eulerian finite element methods. In: Belytschko, T. and Hughes, T. J. R. (eds), *Computational Methods for Transient Analysis*, Elsevier Science, Amsterdam.

Ghoreishy, M. H. R. and Nassehi, V., 1997. Modelling the transient flow of rubber compounds in the dispersive section of an internal mixer with slip–stick boundary conditions. *Adv. Poly. Tech.* **16**, 45–68.

Gresho, P. M., Lee, R. L. and Sani, R. L., 1980. On the time-dependent solution of the incompressible Navier–Stokes equations in two and three dimensions. In: *Recent Advances in Numerical Methods in Fluids*, Ch. 2, Pineridge Press, Swansea, pp. 27–75.

Hughes, T. J.R., 1987. Recent progress in the development and understanding of SUPG methods. *Int. J. Numer. Methods Eng.* **7**, 1261–1275.

Hughes, T. J. R., Franca, L. P. and Balestra, M., 1986. A new finite-element formulation for computational fluid dynamics. 5. Circumventing the Babuska–Brezzi condition - a stable Petrov–Galerkin formulation of the Stokes problem accommodating equal order interpolations. *Comput. Methods Appl. Mech. Eng.* **59**, 85–99.

Hughes, T. J. R., Mallet, M. and Mizukami, A., 1986. A new finite element formulation for computational fluid dynamics: II. Beyond SUPG. *Comput. Methods Appl. Mech. Eng.* **54**, 341–355.

Keunings, R., 1989. Simulation of viscoelastic fluid flow. In: Tucker, C. L. III (ed.), *Computer Modeling for Polymer Processing*, Chapter 9, Hanser Publishers, Munich, pp. 403–469.

Kheshgi, H. S. and Scriven, L. E., 1985. Variable penalty method for finite element analysis of incompressible flow. *Int. J. Numer. Methods Fluids* **5**, 785–803.

Lee, R. L., Gresho, P. M. and Sani, R. L., 1979. Smoothing techniques for certain primitive variable solutions of the Navier–Stokes equations. *Int. J. Numer. Methods Eng.* **14**, 1785–1804.

Luo, X. L. and Tanner, R. I., 1989. A decoupled finite element streamline-upwind scheme for viscoelastic flow problems. *J. Non-Newtonian Fluid Mech.* **31**, 143–162.

Marchal, J. M. and Crochet, M. J., 1987. A new mixed finite element for calculating viscoelastic flow. *J. Non-Newtonian Fluid Mech.* **20**, 77–114.

Morton, K. W., 1996. *Numerical Solution of Convection Diffusion Problems*, Chapman & Hall, London.

Nakazawa, S., Pittman, J. F. T. and Zienkiewicz, O. C., 1982. Numerical solution of flow and heat transfer in polymer melts. In: Gallagher, R. H. *et al.* (eds), *Finite Elements in Fluids*, Vol. 4, Ch. 13, Wiley, Chichester, pp. 251–283.

Nguen, N. and Reynen, J., 1984. A space–time least-squares finite element scheme for advection–diffusion equations. *Comput. Methods Appl. Mech. Eng.* **42**, 331–342.

Nichols, B. D., Hirt, C. W. and Hitchkiss, R. S., 1980. SOLA-VOF: a solution algorithm for transient fluid flow with multiple free surface boundaries. Los Alamos Scientific Laboratories Report No. La-8355, Los Alamos, NM.

Nour-Omid, B., 1987. Lanczos method for heat conduction analysis. *Int. J. Numer. Methods Eng.* **24**, 251–262.

Papanastasiou, T. C., Malamataris, N. and Ellwood, K., 1992. A new outflow boundary condition. *Int. J. Numer. Methods Fluids* **14**, 587–608.

Papanastasiou, T. C., Scriven, L. E. and Macoski, C. W., 1987. A finite element method for liquid with memory. *J. Non-Newtonian Fluid Mech* **22**, 271–288.

Petera, J. and Nassehi, V., 1996. Finite element modelling of free surface viscoelastic flows with particular application to rubber mixing. *Int. J. Numer. Methods Fluids* **23**, 1117–1132.

Petera, J., Nassehi, V. and Pittman, J. F. T., 1993. Petrov–Galerkin methods on isoparametric bilinear and biquadratic elements tested for a scalar convection–diffusion problem. *Int. J. Numer. Methods Heat Fluid Flow* **3**, 205–222.

Petera, J. and Pittman, J. F T., 1994. Isoparametric Hermite elements. *Int. J. Numer. Methods Eng.* **37**, 3489–3519.

Pittman, J. F. T., 1989. Finite elements for field problems. In: Tucker, C. L. III (ed.), *Computer Modeling for Polymer Processing*, Chapter 6, Hanser Publishers, Munich, pp. 237–331.

Pittman, J. F. T. and Nakazawa, S., 1984. Finite element analysis of polymer processing operations. In: Pittman, J. F. T., Zienkiewicz, O. C., Wood, R. D. and Alexander, J. M. (eds), *Numerical Analysis of Forming Processes*, Chapter 6, Wiley, Chichester, pp. 165–218.

Reddy, J. N., 1986. *Applied Functional Analysis and Variational Methods in Engineering*, McGraw-Hill, New York.

Renardy, M., 1997. Imposing 'no' boundary condition at outflow: why does it work? *Int. J. Numer. Methods Fluids* **24**, 413–417.

Sani, R. L. and Gresho, P. M., 1994. Resume and remarks on the Open Boundary Condition Mini-symposium. *Int. J. Numer. Fluids* **18**, 983–1008.

Silliman, W. J. and Scriven, L. E., 1980. Separating flow near a static contact line: slip at a wall and shape of a free surface. *J. Comput. Phys.* **34**, 287–313.

Swarbrick, S. J. and Nassehi, V., 1992a. A new decoupled finite element algorithm for viscoelastic flow. Part 1: numerical algorithm and sample results. *Int. J. Numer. Methods Fluids* **14**, 1367–1376.

Swarbrick, S. J. and Nassehi, V., 1992b. A new decoupled finite element algorithm for viscoelastic flow. Part 2: convergence properties of the algorithm. *Int. J. Numer. Methods Fluids* **14**, 1377–1382.

Tanner, R. I. 2000. *Engineering Rheology*, 2nd edn, Oxford University Press, Oxford.

Taylor, C., Rance, J. and Medwell, J. O., 1985. A note on the imposition of traction boundary conditions when using FEM for solving incompressible flow problems. *Commun. Appl. Numer. Methods* **1**, 113–121.

Thompson, E., 1986. Use of pseudo-concentration to follow creeping flows during transient analysis. *Int. J. Numer. Methods Fluids* **6**, 749–761.

Zienkiewicz, O. C. and Taylor, R. L., 1994. *The Finite Element Method*, 4th edn, Vols 1 and 2, McGraw-Hill, London.

Zienkiewicz, O. C. *et al.*, 1986. The patch test for mixed formulations. *Int. J. Numer. Methods Eng.* **23**, 1873–1883.

Zienkiewicz, O. C. *et al.*, 1985. Iterative method for constrained and mixed approximation, an inexpensive improvement to f.e.m. performance. *Comput. Methods Appl. Mech. Eng.* **51**, 3–29.

Zienkiewicz, O. C. and Wu, J., 1991. Incompressibility without tears – how to avoid restrictions on mixed formulation. *Int. J. Numer. Methods Eng.* **32**, 1189–1203.

4

Working Equations of the Finite Element Schemes

In this chapter derivation of the working equations of various finite element schemes, described in Chapter 3, are explained. These schemes provide the basis for construction of finite element algorithms for simulation of polymeric flow regimes. The working equations of the most commonly used schemes are given in Cartesian, polar and axisymmetric coordinate systems to simplify their incorporation into the finite element program listed in Chapter 7. An example demonstrating the utilization of these equations for modification (or further development) of this program is also included in Chapter 7.

4.1 MODELLING OF STEADY STATE STOKES FLOW OF A GENERALIZED NEWTONIAN FLUID

The majority of polymer flow processes are characterized as low Reynolds number Stokes (i.e. creeping) flow regimes. Therefore in the formulation of finite element models for polymeric flow systems the inertia terms in the equation of motion are usually neglected. In addition, highly viscous polymer flow systems are, in general, dominated by stress and pressure variations and in comparison the body forces acting upon them are small and can be safely ignored.

In this section the governing Stokes flow equations in Cartesian, polar and axisymmetric coordinate systems are presented. The equations given in two-dimensional Cartesian coordinate systems are used to outline the derivation of the elemental stiffness equations (i.e. the working equations) of various finite element schemes.

4.1.1 Governing equations in two-dimensional Cartesian coordinate systems

In the absence of body force the equations of continuity and motion representing Stokes flow in a two-dimensional Cartesian system are written, on the basis of Equations (1.1) and (1.4), as

$$\frac{\partial u}{\partial x} + \frac{\partial v}{\partial y} = 0 \qquad \text{(continuity)} \tag{4.1}$$

where u and v are the x and y components of velocity, respectively, and

$$\begin{cases} -\dfrac{\partial p}{\partial x} + \dfrac{\partial \tau_{xx}}{\partial x} + \dfrac{\partial \tau_{yx}}{\partial y} = 0 \\[2mm] -\dfrac{\partial p}{\partial y} + \dfrac{\partial \tau_{xy}}{\partial x} + \dfrac{\partial \tau_{yy}}{\partial y} = 0 \end{cases} \qquad \text{(motion} \tag{4.2}$$

where p is pressure and τ_{xx} etc. are the components of the extra stress tensor. For a generalized Newtonian fluid using the constitutive equation (1.8)

$$\begin{cases} \tau_{xx} = 2\eta \dfrac{\partial u}{\partial x} \\[2mm] \tau_{yx} = \tau_{xy} = \eta\left(\dfrac{\partial u}{\partial y} + \dfrac{\partial v}{\partial x}\right) \\[2mm] \tau_{yy} = 2\eta \dfrac{\partial v}{\partial y} \end{cases} \tag{4.3}$$

where η is fluid viscosity. Substitution from Equation (4.3) into Equation (4.2) gives

$$\begin{cases} -\dfrac{\partial p}{\partial x} + \dfrac{\partial}{\partial x}\left(2\eta \dfrac{\partial u}{\partial x}\right) + \dfrac{\partial}{\partial y}\left[\eta\left(\dfrac{\partial v}{\partial x} + \dfrac{\partial u}{\partial y}\right)\right] = 0 \\[3mm] -\dfrac{\partial p}{\partial y} + \dfrac{\partial}{\partial x}\left[\eta\left(\dfrac{\partial u}{\partial y} + \dfrac{\partial v}{\partial x}\right)\right] + \dfrac{\partial}{\partial y}\left(2\eta \dfrac{\partial v}{\partial y}\right) = 0 \end{cases} \tag{4.4}$$

Equations (4.1) and (4.4) are the governing flow equations.

4.1.2   Governing equations in two-dimensional polar coordinate systems

Similarly in the absence of body forces the Stokes flow equations for a generalized Newtonian fluid in a two-dimensional (r, θ) coordinate system are written as

$$\frac{\partial v_r}{\partial r} + \frac{v_r}{r} + \frac{1}{r}\frac{\partial v_\theta}{\partial \theta} = 0 \tag{4.5}$$

and

$$
\left\{
\begin{aligned}
&-\frac{\partial p}{\partial r} + \frac{\partial}{\partial r}\left(2\eta\frac{\partial v_r}{\partial r}\right) + \frac{2\eta}{r}\frac{\partial v_r}{\partial r} + \frac{1}{r}\frac{\partial}{\partial\theta}\left[\eta\left(\frac{\partial v_\theta}{\partial r} - \frac{v_\theta}{r} + \frac{1}{r}\frac{\partial v_r}{\partial\theta}\right)\right] - \\
&\qquad\qquad\qquad\qquad\qquad \frac{2\eta}{r}\left(\frac{1}{r}\frac{\partial v_\theta}{\partial\theta} + \frac{v_r}{r}\right) = 0 \\[2mm]
&-\frac{1}{r}\frac{\partial p}{\partial\theta} + \frac{\partial}{\partial r}\left[\eta\left(\frac{\partial v_\theta}{\partial r} - \frac{v_\theta}{r} + \frac{1}{r}\frac{\partial v_r}{\partial\theta}\right)\right] + \frac{2\eta}{r}\left(\frac{\partial v_\theta}{\partial r} - \frac{v_\theta}{r} + \frac{1}{r}\frac{\partial v_r}{\partial\theta}\right) + \\
&\qquad\qquad\qquad\qquad\qquad \frac{2}{r}\frac{\partial}{\partial\theta}\left[\eta\left(\frac{1}{r}\frac{\partial v_\theta}{\partial\theta} + \frac{v_r}{r}\right)\right] = 0
\end{aligned}
\right.
$$

$$(4.6)$$

4.1.4   Governing equations in axisymmetric coordinate systems

In an axisymmetric flow regime all of the field variables remain constant in the circumferential direction around an axis of symmetry. Therefore the governing flow equations in axisymmetric systems can be analytically integrated with respect to this direction to reduce the model to a two-dimensional form. In order to illustrate this procedure we consider the three-dimensional continuity equation for an incompressible fluid written in a cylindrical (r, θ, z) coordinate system as

$$
\frac{1}{r}\frac{\partial}{\partial r}(rv_r) + \frac{1}{r}\frac{\partial}{\partial\theta}(v_\theta) + \frac{\partial}{\partial z}(v_z) = 0 \tag{4.7}
$$

and

$$
\iiint\left[\frac{1}{r}\frac{\partial}{\partial r}(rv_r) + \frac{1}{r}\frac{\partial}{\partial\theta}(v_\theta) + \frac{\partial}{\partial z}(v_z)\right] r\ dr\ d\theta\ dz = 0 \tag{4.8}
$$

In an axisymmetric flow regime there will be no variation in the circumferential (i.e. θ) direction and the second term of the integrand in Equation (4.8) can be eliminated. After integration with respect to θ between the limits of 0–2π Equation (4.8) yields

$$
2\pi\iint\left[\frac{1}{r}\frac{\partial}{\partial r}(rv_r) + \frac{\partial}{\partial z}(v_z)\right] r\ dr\ dz = 0 \tag{4.9}
$$

Therefore the continuity equation for an incompressible axisymmetric flow is written as

$$\frac{\partial v_r}{\partial r} + \frac{v_r}{r} + \frac{\partial v_z}{\partial z} = 0 \tag{4.10}$$

Similarly the components of the equation of motion for an axisymmetric Stokes flow of a generalized Newtonian fluid are written as

$$\begin{cases} -\dfrac{\partial p}{\partial r} + \dfrac{\partial}{\partial r}\left(2\eta\dfrac{\partial v_r}{\partial r}\right) + \dfrac{2\eta}{r}\dfrac{\partial v_r}{\partial r} - \dfrac{2\eta}{r}\dfrac{v_r}{r} + \dfrac{\partial}{\partial z}\left[\eta\left(\dfrac{\partial v_z}{\partial r} + \dfrac{\partial v_r}{\partial z}\right)\right] = 0 \\[4mm] -\dfrac{\partial p}{\partial z} + \dfrac{\partial}{\partial z}\left(2\eta\dfrac{\partial v_z}{\partial z}\right) + \dfrac{\eta}{r}\left(\dfrac{\partial v_z}{\partial r} + \dfrac{\partial v_r}{\partial z}\right) + \dfrac{\partial}{\partial r}\left[\eta\left(\dfrac{\partial v_z}{\partial r} + \dfrac{\partial v_r}{\partial z}\right)\right] = 0 \end{cases} \tag{4.11}$$

4.1.4 Working equations of the U–V–P scheme in Cartesian coordinate systems

Following the procedure described in Chapter 3, Section 1.1 the Galerkin-weighted residual statements corresponding to Equations (4.4) and (4.1) are written as

$$\begin{cases} \displaystyle\int_{\Omega_e} N_i\left[-\frac{\partial\sum_{l=1}^{m}M_l p_l}{\partial x} + 2\frac{\partial}{\partial x}\eta\left(\frac{\partial\sum_{j=1}^{n}N_j u_j}{\partial x}\right) + \frac{\partial}{\partial y}\eta\left(\frac{\partial\sum_{j=1}^{n}N_j v_j}{\partial x} + \frac{\partial\sum_{j=1}^{n}N_j u_j}{\partial y}\right)\right] dx\,dy = 0 \\[6mm] \displaystyle\int_{\Omega_e} N_i\left[-\frac{\partial\sum_{l=1}^{m}M_l p_l}{\partial y} + \frac{\partial}{\partial x}\eta\left(\frac{\partial\sum_{j=1}^{n}N_j u_j}{\partial y} + \frac{\partial\sum_{j=1}^{n}N_j v_j}{\partial x}\right) + 2\frac{\partial}{\partial y}\eta\left(\frac{\partial\sum_{j=1}^{n}N_j v_j}{\partial y}\right)\right] dx\,dy = 0 \\[6mm] \displaystyle\int_{\Omega_e} -M_l\left(\frac{\partial\sum_{j=1}^{n}N_j u_j}{\partial x} + \frac{\partial\sum_{j=1}^{n}N_j v_j}{\partial y}\right) dx\,dy = 0 \end{cases} \tag{4.12}$$

In Equation (4.12) the discretization of velocity and pressure is based on different shape functions (i.e. $N_j\ j = 1,n$ and $M_l\ l = 1,m$ where, in general, $m<n$). The weight function used in the continuity equation is selected as $-M_l$ to retain the symmetry of the discretized equations. After application of Green's theorem to the second-order velocity derivatives (to reduce inter-element continuity requirement) and the pressure terms (to maintain the consistency of the formulation) and algebraic manipulations the working equations of the U–V–P scheme are obtained as

$$\begin{bmatrix} A_{ij}^{11} & A_{ij}^{12} & A_{il}^{13} \\ A_{ij}^{21} & A_{ij}^{22} & A_{il}^{23} \\ A_{lj}^{31} & A_{lj}^{32} & A_{ll}^{33} \end{bmatrix} \begin{Bmatrix} u_j \\ v_j \\ p_l \end{Bmatrix} = \begin{Bmatrix} B_j^1 \\ B_j^2 \\ B_l^3 \end{Bmatrix} \tag{4.13}$$

where

$$A_{ij}^{11} = \int_{\Omega_e} \left(2\eta \frac{\partial N_i}{\partial x} \frac{\partial N_j}{\partial x} + \eta \frac{\partial N_i}{\partial y} \frac{\partial N_j}{\partial y} \right) \mathrm{d}x\,\mathrm{d}y \tag{4.14}$$

$$A_{ij}^{12} = \int_{\Omega_e} \left(\eta \frac{\partial N_i}{\partial y} \frac{\partial N_j}{\partial x} \right) \mathrm{d}x\,\mathrm{d}y \tag{4.15}$$

$$A_{il}^{13} = -\int_{\Omega_e} \left(M_l \frac{\partial N_i}{\partial x} \right) \mathrm{d}x\,\mathrm{d}y \tag{4.16}$$

$$A_{ij}^{21} = \int_{\Omega_e} \left(\eta \frac{\partial N_i}{\partial x} \frac{\partial N_j}{\partial y} \right) \mathrm{d}x\,\mathrm{d}y \tag{4.17}$$

$$A_{ij}^{22} = \int_{\Omega_e} \left(\eta \frac{\partial N_i}{\partial x} \frac{\partial N_j}{\partial x} + 2\eta \frac{\partial N_i}{\partial y} \frac{\partial N_j}{\partial y} \right) \mathrm{d}x\,\mathrm{d}y \tag{4.18}$$

$$A_{il}^{23} = -\int_{\Omega_e} \left(M_l \frac{\partial N_i}{\partial y} \right) \mathrm{d}x\,\mathrm{d}y \tag{4.19}$$

$$A_{lj}^{31} = -\int_{\Omega_e} \left(M_l \frac{\partial N_j}{\partial x} \right) \mathrm{d}x\,\mathrm{d}y \tag{4.20}$$

$$A_{lj}^{32} = -\int_{\Omega_e} \left(M_l \frac{\partial N_j}{\partial y} \right) \mathrm{d}x\,\mathrm{d}y \tag{4.21}$$

$$A_{ll}^{33} = 0 \tag{4.22}$$

$$B_j^1 = \int_{\Gamma_e} N_i \left[\left(-p^e + 2\eta \frac{\partial u^e}{\partial x} \right) n_x + \eta \left(\frac{\partial u^e}{\partial y} + \frac{\partial v^e}{\partial x} \right) n_y \right] \mathrm{d}\Gamma_e \tag{4.23}$$

$$B_j^2 = \int_{\Gamma_e} N_i \left[\eta \left(\frac{\partial u^e}{\partial y} + \frac{\partial v^e}{\partial x} \right) n_x + \left(-p^e + 2\eta \frac{\partial v^e}{\partial y} \right) n_y \right] \mathrm{d}\Gamma_e \tag{4.24}$$

$$B_l^3 = 0 \tag{4.25}$$

where Γ_e is the element boundary and n_x and n_y are the components of unit outward vector normal to Γ_e. A superscript e is used to indicate elemental discretization (e.g. $p^e = \sum_{l=1}^{m} M_l p_l$) throughout this chapter.

4.1.5   Working equations of the U–V–P scheme in polar coordinate systems

Using a procedure similar to the derivation of Equation (4.13) the working equations of the U–V–P scheme for steady-state Stokes flow in a polar (r, θ) coordinate system are obtained on the basis of Equations (4.5) and (4.6) as

$$
\begin{bmatrix} A_{ij}^{11} & A_{ij}^{12} & A_{il}^{13} \\ A_{ij}^{21} & A_{ij}^{22} & A_{il}^{23} \\ A_{lj}^{31} & A_{lj}^{32} & A_{ll}^{33} \end{bmatrix} \begin{Bmatrix} v_{rj} \\ v_{\theta j} \\ p_l \end{Bmatrix} = \begin{Bmatrix} B_j^1 \\ B_j^2 \\ B_l^3 \end{Bmatrix}
\tag{4.26}
$$

where

$$
A_{ij}^{11} = \int_{\Omega_e} \left(2\eta \frac{\partial N_i}{\partial r} \frac{\partial N_j}{\partial r} + \frac{2\eta}{r^2} N_i N_j + \frac{\eta}{r^2} \frac{\partial N_i}{\partial \theta} \frac{\partial N_j}{\partial \theta} \right) r\, dr\, d\theta
\tag{4.27}
$$

$$
A_{ij}^{12} = \int_{\Omega_e} \left(\frac{2\eta}{r^2} N_i \frac{\partial N_j}{\partial \theta} + \frac{\eta}{r} \frac{\partial N_i}{\partial \theta} \frac{\partial N_j}{\partial r} - \frac{\eta}{r^2} \frac{\partial N_i}{\partial \theta} N_j \right) r\, dr\, d\theta
\tag{4.28}
$$

$$
A_{il}^{13} = - \int_{\Omega_e} M_l \left(\frac{\partial N_i}{\partial r} + \frac{N_i}{r} \right) r\, dr\, d\theta
\tag{4.29}
$$

$$
A_{ij}^{21} = \int_{\Omega_e} \left(\frac{\eta}{r} \frac{\partial N_i}{\partial r} \frac{\partial N_j}{\partial \theta} - \frac{\eta}{r} \frac{N_i}{r} \frac{\partial N_j}{\partial \theta} + \frac{2\eta}{r} \frac{\partial N_i}{\partial \theta} \frac{N_j}{r} \right) r\, dr\, d\theta
\tag{4.30}
$$

$$
A_{ij}^{22} = \int_{\Omega_e} \left(\eta \frac{\partial N_i}{\partial r} \frac{\partial N_j}{\partial r} - \eta \frac{N_i}{r} \frac{\partial N_j}{\partial r} - \eta \frac{\partial N_i}{\partial r} \frac{N_j}{r} + \eta \frac{N_i}{r} \frac{N_j}{r} + \frac{2\eta}{r^2} \frac{\partial N_i}{\partial \theta} \frac{\partial N_j}{\partial \theta} \right) r\, dr\, d\theta
\tag{4.31}
$$

$$
A_{il}^{23} = - \int_{\Omega_e} \left(\frac{1}{r} M_l \frac{\partial N_i}{\partial \theta} \right) r\, dr\, d\theta
\tag{4.32}
$$

$$
A_{lj}^{31} = - \int_{\Omega_e} M_l \left(\frac{\partial N_j}{\partial r} + \frac{N_j}{r} \right) r\, dr\, d\theta
\tag{4.33}
$$

$$
A_{lj}^{32} = - \int_{\Omega_e} \left(M_l \frac{1}{r} \frac{\partial N_j}{\partial \theta} \right) r\, dr\, d\theta
\tag{4.34}
$$

$$
A_{ll}^{33} = 0
\tag{4.35}
$$

$$B_j^1 = \int_{\Gamma_e} N_i \left[\left(-p^e + 2\eta \frac{\partial v_r^e}{\partial r} \right) n_r + \frac{\eta}{r} \left(\frac{\partial v_\theta^e}{\partial r} - \frac{v_\theta^e}{r} + \frac{1}{r} \frac{\partial v_r^e}{\partial \theta} \right) n_\theta \right] r \, d\Gamma_e \qquad (4.36)$$

$$B_j^2 = \int_{\Gamma_e} N_i \left[\eta \left(\frac{\partial v_\theta^e}{\partial r} - \frac{v_\theta^e}{r} + \frac{1}{r} \frac{\partial v_r^e}{\partial \theta} \right) n_r + \left(-\frac{1}{r} p^e + \frac{2\eta}{r^2} \frac{\partial v_\theta^e}{\partial \theta} + \frac{2\eta}{r^2} v_r^e \right) n_\theta \right] r \, d\Gamma_e \qquad (4.37)$$

$$B_l^3 = 0 \qquad (4.38)$$

4.1.6 Working equations of the U–V–P scheme in axisymmetric coordinate systems

Using a procedure similar to the formulation of two-dimensional forms the working equations of the U–V–P scheme in axisymmetric coordinate systems are derived on the basis of Equations (4.10) and (4.11) as

$$\begin{bmatrix} A_{ij}^{11} & A_{ij}^{12} & A_{il}^{13} \\ A_{ij}^{21} & A_{ij}^{22} & A_{il}^{23} \\ A_{lj}^{31} & A_{lj}^{32} & A_{ll}^{33} \end{bmatrix} \begin{Bmatrix} v_{rj} \\ v_{zj} \\ p_l \end{Bmatrix} = \begin{Bmatrix} B_j^1 \\ B_j^2 \\ B_l^3 \end{Bmatrix} \qquad (4.39)$$

where

$$A_{ij}^{11} = \int_{\Omega_e} \left(2\eta \frac{\partial N_i}{\partial r} \frac{\partial N_j}{\partial r} + \frac{2\eta}{r^2} N_i N_j + \eta \frac{\partial N_i}{\partial z} \frac{\partial N_j}{\partial z} \right) r \, dr \, dz \qquad (4.40)$$

$$A_{ij}^{12} = \int_{\Omega_e} \left(\eta \frac{\partial N_i}{\partial z} \frac{\partial N_j}{\partial r} \right) r \, dr \, dz \qquad (4.41)$$

$$A_{il}^{13} = - \int_{\Omega_e} M_l \left(\frac{\partial N_i}{\partial r} + \frac{N_i}{r} \right) r \, dr \, dz \qquad (4.42)$$

$$A_{ij}^{21} = \int_{\Omega_e} \left(\eta \frac{\partial N_i}{\partial r} \frac{\partial N_j}{\partial z} \right) r \, dr \, dz \qquad (4.43)$$

$$A_{ij}^{22} = \int_{\Omega_e} \left(\eta \frac{\partial N_i}{\partial r} \frac{\partial N_j}{\partial r} + 2\eta \frac{\partial N_i}{\partial z} \frac{\partial N_j}{\partial z} \right) r \, dr \, dz \qquad (4.44)$$

$$A_{il}^{23} = - \int_{\Omega_e} \left(M_l \frac{\partial N_i}{\partial z} \right) r \, dr \, dz \qquad (4.45)$$

$$A_{lj}^{31} = - \int_{\Omega_e} M_l \left(\frac{\partial N_j}{\partial r} + \frac{N_j}{r} \right) r \, dr \, dz \tag{4.46}$$

$$A_{lj}^{32} = - \int_{\Omega_e} \left(M_l \frac{\partial N_j}{\partial z} \right) r \, dr \, dz \tag{4.47}$$

$$A_{ll}^{33} = 0 \tag{4.48}$$

$$B_j^1 = \int_{\Gamma_e} N_i \left[\left(-p^e + 2\eta \frac{\partial v_r^e}{\partial r} \right) n_r + \frac{\eta}{r} \left(\frac{\partial v_r^e}{\partial z} + \frac{1}{r} \frac{\partial v_z^e}{\partial r} \right) n_z \right] r \, d\Gamma_e \tag{4.49}$$

$$B_j^2 = \int_{\Gamma_e} N_i \left[\eta \left(\frac{\partial v_z^e}{\partial r} + \frac{\partial v_r^e}{\partial z} \right) n_r + \left(-p^e + 2\eta \frac{\partial v_z^e}{\partial z} \right) n_z \right] r \, d\Gamma_e \tag{4.50}$$

$$B_l^3 = 0 \tag{4.51}$$

4.1.7 Working equations of the continuous penalty scheme in Cartesian coordinate systems

In the continuous penalty method prior to the discretization of the flow equations the pressure term in the equation of motion is substituted by the penalty relationship, given as Equation (3.6). Therefore using Equations (4.4) and (4.1), we have

$$\begin{cases} -\dfrac{\partial}{\partial x} \left[-\lambda \left(\dfrac{\partial u}{\partial x} + \dfrac{\partial v}{\partial y} \right) \right] + \dfrac{\partial}{\partial x} \left(2\eta \dfrac{\partial u}{\partial x} \right) + \dfrac{\partial}{\partial y} \left[\eta \left(\dfrac{\partial v}{\partial x} + \dfrac{\partial u}{\partial y} \right) \right] = 0 \\[4mm] -\dfrac{\partial}{\partial y} \left[-\lambda \left(\dfrac{\partial u}{\partial x} + \dfrac{\partial v}{\partial y} \right) \right] + \dfrac{\partial}{\partial x} \left[\eta \left(\dfrac{\partial u}{\partial y} + \dfrac{\partial v}{\partial x} \right) \right] + \dfrac{\partial}{\partial y} \left[2\eta \left(\dfrac{\partial v}{\partial y} \right) \right] = 0 \end{cases} \tag{4.52}$$

Following the procedure described in the continuous penalty technique subsection in Chapter 3 the Galerkin-weighted residual statements corresponding to Equation (4.52) are written as

$$
\begin{cases}
\displaystyle\int_{\Omega_e} N_i \left[\frac{\partial}{\partial x} \lambda \left(\frac{\partial \sum_{j=1}^{n} N_j u_j}{\partial x} + \frac{\partial \sum_{j=1}^{n} N_j v_j}{\partial y} \right) + \frac{\partial}{\partial x} 2\eta \frac{\partial \sum_{j=1}^{n} N_j u_j}{\partial x} + \right. \\
\qquad\qquad \left. \frac{\partial}{\partial y} \left[\eta \left(\frac{\partial \sum_{j=1}^{n} N_j v_j}{\partial x} + \frac{\partial \sum_{j=1}^{n} N_j u_j}{\partial y} \right) \right] \right] dx\,dy = 0 \\[2em]
\displaystyle\int_{\Omega_e} N_i \left[\frac{\partial}{\partial y} \lambda \left(\frac{\partial \sum_{j=1}^{n} N_j u_j}{\partial x} + \frac{\partial \sum_{j=1}^{n} N_j v_j}{\partial y} \right) + \frac{\partial}{\partial y} \left[\eta \left(\frac{\partial \sum_{j=1}^{n} N_j u_j}{\partial y} + \frac{\partial \sum_{j=1}^{n} N_j v_j}{\partial x} \right) \right] + \right. \\
\qquad\qquad \left. \frac{\partial}{\partial x} 2\eta \frac{\partial \sum_{j=1}^{n} N_j v_j}{\partial y} \right] dx\,dy = 0
\end{cases}
$$

$$(4.53)$$

After application of Green's theorem to the second-order velocity derivatives (to reduce inter-element continuity requirement) and algebraic manipulations the working equations of the continuous penalty scheme are obtained as

$$
\begin{bmatrix} A_{ij}^{11} & A_{ij}^{12} \\ A_{ij}^{21} & A_{ij}^{22} \end{bmatrix} \begin{Bmatrix} u_j \\ v_j \end{Bmatrix} = \begin{Bmatrix} B_j^1 \\ B_j^2 \end{Bmatrix}
\tag{4.54}
$$

where

$$
A_{ij}^{11} = \int_{\Omega_e} \left[(\lambda + 2\eta) \left(\frac{\partial N_i}{\partial x} \frac{\partial N_j}{\partial x} \right) + \eta \frac{\partial N_i}{\partial y} \frac{\partial N_j}{\partial y} \right] dx\,dy
\tag{4.55}
$$

$$
A_{ij}^{12} = \int_{\Omega_e} \left(\lambda \frac{\partial N_i}{\partial x} \frac{\partial N_j}{\partial y} + \eta \frac{\partial N_i}{\partial y} \frac{\partial N_j}{\partial x} \right) dx\,dy
\tag{4.56}
$$

$$
A_{ij}^{21} = \int_{\Omega_e} \left(\lambda \frac{\partial N_i}{\partial y} \frac{\partial N_j}{\partial x} + \eta \frac{\partial N_i}{\partial x} \frac{\partial N_j}{\partial y} \right) dx\,dy
\tag{4.57}
$$

$$
A_{ij}^{22} = \int_{\Omega_e} \left[(\lambda + 2\eta) \left(\frac{\partial N_i}{\partial y} \frac{\partial N_j}{\partial y} \right) + \eta \frac{\partial N_i}{\partial x} \frac{\partial N_j}{\partial x} \right] dx\,dy
\tag{4.58}
$$

$$B_j^1 = \int_{\Gamma_e} N_i \left\{ \left[\lambda \left(\frac{\partial u^e}{\partial x} + \frac{\partial v^e}{\partial y} \right) + 2\eta \frac{\partial u^e}{\partial x} \right] n_x + \eta \left(\frac{\partial v^e}{\partial x} + \frac{\partial u^e}{\partial y} \right) n_y \right\} d\Gamma_e \qquad (4.59)$$

$$B_j^2 = \int_{\Gamma_e} N_i \left\{ \eta \left(\frac{\partial v^e}{\partial x} + \frac{\partial u^e}{\partial y} \right) n_x + \left[\lambda \left(\frac{\partial u^e}{\partial x} + \frac{\partial v^e}{\partial y} \right) + 2\eta \frac{\partial v^e}{\partial y} \right] n_y \right\} d\Gamma_e \qquad (4.60)$$

As it can be seen the working equations of the penalty scheme are more compact than their counterparts obtained for the U–V–P method.

In some applications it may be necessary to prescribe a pressure datum at a node at the domain boundary. Although pressure has been eliminated from the working equations in the penalty scheme it can be reintroduced through the penalty terms appearing in the boundary line integrals.

4.1.8 Working equations of the continuous penalty scheme in polar coordinate systems

After the substitution of pressure via the penalty relationship the flow equations in a polar coordinate system are written as

$$
\left\{
\begin{aligned}
& -\frac{\partial}{\partial r} \left[-\lambda \left(\frac{\partial v_r}{\partial r} + \frac{v_r}{r} + \frac{1}{r} \frac{\partial v_\theta}{\partial \theta} \right) \right] + \frac{\partial}{\partial r} \left(2\eta \frac{\partial v_r}{\partial r} \right) + \frac{2\eta}{r} \frac{\partial v_r}{\partial r} + \\
& \qquad\qquad \frac{1}{r} \frac{\partial}{\partial \theta} \left[\eta \left(\frac{\partial v_\theta}{\partial r} - \frac{v_\theta}{r} + \frac{1}{r} \frac{\partial v_r}{\partial \theta} \right) \right] - \frac{2\eta}{r} \left(\frac{1}{r} \frac{\partial v_\theta}{\partial \theta} + \frac{v_r}{r} \right) = 0 \\
& -\frac{1}{r} \frac{\partial}{\partial \theta} \left[-\lambda \left(\frac{\partial v_r}{\partial r} + \frac{v_r}{r} + \frac{1}{r} \frac{\partial v_\theta}{\partial \theta} \right) \right] + \frac{\partial}{\partial r} \left[\eta \left(\frac{\partial v_\theta}{\partial r} - \frac{v_\theta}{r} + \frac{1}{r} \frac{\partial v_r}{\partial \theta} \right) \right] + \\
& \qquad\qquad \frac{2\eta}{r} \left(\frac{\partial v_\theta}{\partial r} - \frac{v_\theta}{r} + \frac{1}{r} \frac{\partial v_r}{\partial \theta} \right) + \frac{2}{r} \frac{\partial}{\partial \theta} \left[\eta \left(\frac{1}{r} \frac{\partial v_\theta}{\partial \theta} + \frac{v_r}{r} \right) \right] = 0
\end{aligned}
\right. \qquad (4.61)
$$

Using a procedure similar to the derivation of Equation (4.53) the working equations of the continuous penalty scheme for steady-state Stokes flow in a polar (r, θ) coordinate system are obtained as

$$
\begin{bmatrix} A_{ij}^{11} & A_{ij}^{12} \\ A_{ij}^{21} & A_{ij}^{22} \end{bmatrix}
\begin{Bmatrix} v_{r\,j} \\ v_{\theta\,j} \end{Bmatrix}
=
\begin{Bmatrix} B_j^1 \\ B_j^2 \end{Bmatrix}
\qquad (4.62)
$$

where

$$
A_{ij}^{11} = \int_{\Omega_e} \left[(\lambda + 2\eta) \left(\frac{\partial N_i}{\partial r} \frac{\partial N_j}{\partial r} + \frac{N_i}{r} \frac{N_j}{r} \right) + \lambda \left(\frac{N_i}{r} \frac{\partial N_j}{\partial r} + \frac{\partial N_i}{\partial r} \frac{N_j}{r} \right) + \right.
$$

$$
\left. \frac{\eta}{r^2} \frac{\partial N_i}{\partial \theta} \frac{\partial N_j}{\partial \theta} \right] r \, dr \, d\theta \qquad (4.63)
$$

$$
A_{ij}^{12} = \int_{\Omega_e} \left[\frac{\lambda}{r} \left(\frac{\partial N_i}{\partial r} \frac{\partial N_j}{\partial \theta} + \frac{N_i}{r} \frac{\partial N_j}{\partial \theta} \right) + \frac{\eta}{r} \left(\frac{\partial N_i}{\partial \theta} \frac{\partial N_j}{\partial r} - \frac{\partial N_i}{\partial \theta} \frac{N_j}{r} + \right. \right.
$$

$$
\left. \left. 2 \frac{N_i}{r} \frac{\partial N_j}{\partial \theta} \right) \right] r \, dr \, d\theta \qquad (4.64)
$$

$$
A_{ij}^{21} = \int_{\Omega_e} \left[\frac{\lambda}{r} \left(\frac{\partial N_i}{\partial \theta} \frac{\partial N_j}{\partial r} + \frac{N_j}{r} \frac{\partial N_i}{\partial \theta} \right) + \frac{\eta}{r} \left(\frac{\partial N_i}{\partial r} \frac{\partial N_j}{\partial \theta} - \frac{\partial N_j}{\partial \theta} \frac{N_i}{r} + \right. \right.
$$

$$
\left. \left. 2 \frac{N_j}{r} \frac{\partial N_i}{\partial \theta} \right) \right] r \, dr \, d\theta \qquad (4.65)
$$

$$
A_{ij}^{22} = \int_{\Omega_e} \left[\frac{\lambda}{r^2} \frac{\partial N_i}{\partial \theta} \frac{\partial N_j}{\partial \theta} + \eta \left(\frac{\partial N_i}{\partial r} \frac{\partial N_j}{\partial r} - \frac{N_i}{r} \frac{\partial N_j}{\partial r} - \frac{\partial N_i}{\partial r} \frac{N_j}{r} + \frac{N_i}{r} \frac{N_j}{r} + \right. \right.
$$

$$
\left. \left. \frac{2}{r^2} \frac{\partial N_i}{\partial \theta} \frac{\partial N_j}{\partial \theta} \right) \right] r \, dr \, d\theta \qquad (4.66)
$$

$$
B_j^1 = \int_{\Gamma_e} N_i \left\{ \left[\lambda \left(\frac{\partial v_r^e}{\partial r} + \frac{v_r^e}{r} + \frac{1}{r} \frac{\partial v_\theta^e}{\partial \theta} \right) + 2\eta \frac{\partial v_r^e}{\partial r} \right] n_r + \right.
$$

$$
\left. \frac{\eta}{r} \left(\frac{\partial v_\theta^e}{\partial r} - \frac{v_\theta^e}{r} + \frac{1}{r} \frac{\partial v_r^e}{\partial \theta} \right) n_\theta \right\} r \, d\Gamma_e \qquad (4.67)
$$

$$
B_j^2 = \int_{\Gamma_e} N_i \left\{ \eta \left(\frac{\partial v_\theta^e}{\partial r} - \frac{v_\theta^e}{r} + \frac{1}{r} \frac{\partial v_r^e}{\partial \theta} \right) n_r + \left[\frac{\lambda}{r} \left(\frac{\partial v_r^e}{\partial r} + \frac{v_r^e}{r} + \frac{1}{r} \frac{\partial v_\theta^e}{\partial \theta} \right) + \right. \right.
$$

$$
\left. \left. \frac{2\eta}{r^2} \left(\frac{\partial v_\theta^e}{\partial \theta} + v_r^e \right) \right] n_\theta \right\} r \, d\Gamma_e \qquad (4.68)
$$

4.1.9 Working equations of the continuous penalty scheme in axisymmetric coordinate systems

After the substitution of pressure via the penalty relationship the flow equations in an axisymmetric coordinate system are written as

$$\left\{ \begin{array}{l} -\dfrac{\partial}{\partial r}\left[-\lambda\left(\dfrac{\partial v_r}{\partial r}+\dfrac{v_r}{r}+\dfrac{\partial v_z}{\partial z}\right)\right]+\dfrac{\partial}{\partial r}\left(2\eta\dfrac{\partial v_r}{\partial r}\right)+\dfrac{2\eta}{r}\dfrac{\partial v_r}{\partial r}-\dfrac{2\eta v_r}{r^2}+ \\[3mm] \qquad\qquad\qquad\qquad \dfrac{\partial}{\partial z}\left[\eta\left(\dfrac{\partial v_z}{\partial r}+\dfrac{\partial v_r}{\partial z}\right)\right]=0 \\[5mm] -\dfrac{\partial}{\partial z}\left[-\lambda\left(\dfrac{\partial v_r}{\partial r}+\dfrac{v_r}{r}+\dfrac{\partial v_z}{\partial z}\right)\right]+\dfrac{\partial}{\partial r}\left[\eta\left(\dfrac{\partial v_z}{\partial r}+\dfrac{\partial v_r}{\partial z}\right)\right]+\dfrac{\eta}{r}\left(\dfrac{\partial v_z}{\partial r}+\dfrac{\partial v_r}{\partial z}\right)+ \\[3mm] \qquad\qquad\qquad\qquad \dfrac{\partial}{\partial z}\left(2\eta\dfrac{\partial v_z}{\partial z}\right)=0 \end{array} \right. \tag{4.69}$$

Using a procedure similar to the derivation of Equation (4.53) the working equations of the continuous penalty scheme for steady-state Stokes flow in an axisymmetric coordinate system are obtained as

$$\begin{bmatrix} A_{ij}^{11} & A_{ij}^{12} \\ A_{ij}^{21} & A_{ij}^{22} \end{bmatrix}\begin{Bmatrix} v_{rj} \\ v_{zj} \end{Bmatrix}=\begin{Bmatrix} B_j^1 \\ B_j^2 \end{Bmatrix} \tag{4.70}$$

where

$$A_{ij}^{11}=\int_{\Omega_e}\left[\lambda\left(\frac{\partial N_i}{\partial r}\frac{\partial N_j}{\partial r}+\frac{N_i}{r}\frac{N_j}{r}+\frac{N_i}{r}\frac{\partial N_j}{\partial r}+\frac{\partial N_i}{\partial r}\frac{N_j}{r}\right)+\right.$$
$$\left.\eta\left(\frac{2N_iN_j}{r^2}+2\frac{\partial N_i}{\partial r}\frac{\partial N_j}{\partial r}+\frac{\partial N_i}{\partial z}\frac{\partial N_j}{\partial z}\right)\right]r\ dr\ dz \tag{4.71}$$

$$A_{ij}^{12}=\int_{\Omega_e}\left[\lambda\left(\frac{\partial N_i}{\partial r}\frac{\partial N_j}{\partial z}+\frac{N_i}{r}\frac{\partial N_j}{\partial z}\right)+\eta\frac{\partial N_i}{\partial z}\frac{\partial N_j}{\partial r}\right]r\ dr\ dz \tag{4.72}$$

$$A_{ij}^{21}=\int_{\Omega_e}\left[\lambda\left(\frac{\partial N_i}{\partial z}\frac{\partial N_j}{\partial r}+\frac{N_j}{r}\frac{\partial N_i}{\partial z}\right)+\eta\frac{\partial N_i}{\partial r}\frac{\partial N_j}{\partial z}\right]r\ dr\ dz \tag{4.73}$$

$$A_{ij}^{22}=\int_{\Omega_e}\left[\lambda\frac{\partial N_i}{\partial z}\frac{\partial N_j}{\partial z}+\eta\left(\frac{\partial N_i}{\partial r}\frac{\partial N_j}{\partial r}+2\frac{\partial N_i}{\partial z}\frac{\partial N_j}{\partial z}\right)\right]r\ dr\ dz \tag{4.74}$$

$$B_j^1=\int_{\Gamma_e}N_i\left\{\left[\lambda\left(\frac{\partial v_r^e}{\partial r}+\frac{v_r^e}{r}+\frac{\partial v_z^e}{\partial z}\right)+2\eta\frac{\partial v_r^e}{\partial z}\right]n_r+\eta\left(\frac{\partial v_r^e}{\partial z}+\frac{\partial v_z^e}{\partial r}\right)n_z\right\}r\ d\Gamma_e \tag{4.75}$$

$$B_j^2 = \int_{\Gamma_e} N_i \left\{ \eta \left(\frac{\partial v_z^e}{\partial r} + \frac{\partial v_r^e}{\partial z} \right) n_r + \left[\lambda \left(\frac{\partial v_r^e}{\partial r} + \frac{v_r^e}{r} + \frac{\partial v_z^e}{\partial z} \right) + 2\eta \frac{\partial v_z^e}{\partial z} \right] n_z \right\} r \, d\Gamma_e \qquad (4.76)$$

4.1.10 Working equations of the discrete penalty scheme in Cartesian coordinate systems

As described in the discrete penalty technique subsection in Chapter 3 in the discrete penalty method components of the equation of motion and the penalty relationship (i.e. the modified equation of continuity) are discretized separately and then used to eliminate the pressure term from the equation of motion. In order to illustrate this procedure we consider the following penalty relationship

$$p = -\lambda \left(\frac{\partial u}{\partial x} + \frac{\partial v}{\partial y} \right) \qquad (4.77)$$

Discretization of Equation (4.77) gives

$$\int_{\Omega_e} M_l \left(\sum_{k=1}^{m} M_k \, p_k \right) dxdy = - \int_{\Omega_e} M_l \lambda \left(\frac{\partial \sum_{j=1}^{n} N_j u_j}{\partial x} + \frac{\partial \sum_{j=1}^{n} N_j v_j}{\partial y} \right) dxdy \qquad (4.78)$$

where M_l is a weight function, identical to the shape functions M_k, $k = 1,m$ which are used for the discretization of the pressure, and N_j, $j = 1,n$ are the shape functions used to discretize the velocity components. Using matrix notation Equation (4.78) is written as

$$\left[\int_{\Omega_e} M_l M_k dxdy \right] \{p_k\} = \left[- \int_{\Omega_e} \lambda M_l \frac{\partial N_j}{\partial x} dxdy \right] \{u_j\} +$$

$$\left[- \int_{\Omega_e} \lambda M_l \frac{\partial N_j}{\partial y} dxdy \right] \{v_j\} \qquad (4.79)$$

Therefore

$$\{p_k\} = \left[\int_{\Omega_e} M_l M_k \, dxdy \right]^{-1} \left[- \int_{\Omega_e} \lambda M_l \frac{\partial N_j}{\partial x} \, dxdy \right] \{u_j\} +$$

$$\left[\int_{\Omega_e} M_l M_k \, dxdy \right]^{-1} \left[- \int_{\Omega_e} \lambda M_l \frac{\partial N_j}{\partial y} \, dxdy \right] \{v_j\} \qquad (4.80)$$

After substitution of pressure in the equation of motion using Equation (4.80) and the application of Green's theorem to the second-order derivatives the working equations of the discrete penalty method are obtained as

$$
\begin{bmatrix} A_{ij}^{11} & A_{ij}^{12} \\ A_{ij}^{21} & A_{ij}^{22} \end{bmatrix} \begin{Bmatrix} u_j \\ v_j \end{Bmatrix} = \begin{Bmatrix} B_j^1 \\ B_j^2 \end{Bmatrix}
\tag{4.81}
$$

where

$$
A_{ij}^{11} = \int_{\Omega_e} \eta \left(2 \frac{\partial N_i}{\partial x} \frac{\partial N_j}{\partial x} + \frac{\partial N_i}{\partial y} \frac{\partial N_j}{\partial y} \right) dxdy +
$$
$$
\left(\int_{\Omega_e} -M_k \frac{\partial N_i}{\partial x} dxdy \right) \left(\int_{\Omega_e} M_l M_k \, dxdy \right)^{-1} \left(\int_{\Omega_e} -\lambda M_l \frac{\partial N_j}{\partial x} dxdy \right)
\tag{4.82}
$$

$$
A_{ij}^{12} = \int_{\Omega_e} \eta \frac{\partial N_i}{\partial y} \frac{\partial N_j}{\partial x} \, dxdy +
$$
$$
\left(\int_{\Omega_e} -M_k \frac{\partial N_i}{\partial x} dxdy \right) \left(\int_{\Omega_e} M_l M_k \, dxdy \right)^{-1} \left(\int_{\Omega_e} -\lambda M_l \frac{\partial N_j}{\partial y} dxdy \right)
\tag{4.83}
$$

$$
A_{ij}^{21} = \int_{\Omega_e} \eta \frac{\partial N_i}{\partial x} \frac{\partial N_j}{\partial y} \, dxdy +
$$
$$
\left(\int_{\Omega_e} -M_k \frac{\partial N_i}{\partial y} dxdy \right) \left(\int_{\Omega_e} M_l M_k \, dxdy \right)^{-1} \left(\int_{\Omega_e} -\lambda M_l \frac{\partial N_j}{\partial x} dxdy \right)
\tag{4.84}
$$

$$
A_{ij}^{22} = \int_{\Omega_e} \eta \left(2 \frac{\partial N_i}{\partial y} \frac{\partial N_j}{\partial y} + \frac{\partial N_i}{\partial x} \frac{\partial N_j}{\partial x} \right) dxdy +
$$
$$
\left(\int_{\Omega_e} -M_k \frac{\partial N_i}{\partial y} dxdy \right) \left(\int_{\Omega_e} M_l M_k \, dxdy \right)^{-1} \left(\int_{\Omega_e} -\lambda M_l \frac{\partial N_j}{\partial y} dxdy \right)
\tag{4.85}
$$

$$
B_j^1 = \int_{\Gamma_e} N_i \left[\left(-p^e + 2\eta \frac{\partial u^e}{\partial x} \right) n_x + \eta \left(\frac{\partial u^e}{\partial y} + \frac{\partial v^e}{\partial x} \right) n_y \right] d\Gamma_e
\tag{4.86}
$$

$$B_j^2 = \int_{\Gamma_e} N_i \left[\eta \left(\frac{\partial u^e}{\partial y} + \frac{\partial v^e}{\partial x} \right) n_x + \left(-p^e + 2\eta \frac{\partial v^e}{\partial y} \right) n_y \right] d\Gamma_e \qquad (4.87)$$

In conjunction with the discrete penalty schemes elements belonging to the Crouzeix–Raviart group are usually used. As explained in Chapter 2, these elements generate discontinuous pressure variation across the inter-element boundaries in a mesh and, hence, the required matrix inversion in the working equations of this scheme can be carried out at the elemental level with minimum computational cost.

4.1.11   Working equations of the least-squares scheme in Cartesian coordinate systems

Following the procedure described in Chapter 3, Section 1.6, after the discretization of Equations (4.1) and (4.4) the generated residuals are used to formulate a functional as

$$I = \int_{\Omega_e} \left[(f_{\text{res}}^u)^2 + (f_{\text{res}}^v)^2 + k(f_{\text{res}}^c)^2 \right] dx dy \qquad (4.88)$$

where f_{res}^u, f_{res}^v and f_{res}^c represent the obtained residuals corresponding to u and v components of the equation of motion and the equation of continuity, respectively. The constant k in functional (4.88) is used to make the functional dimensionally consistent. The magnitude of this constant depends on fluid viscosity and usually a large number (comparable to the parameter used in the penalty schemes) is selected. Utilizing equal order elemental interpolations for the velocity and pressure, functional (4.88) is written as

$$I = \int_{\Omega_e} \left\{ \begin{array}{l} \left[-\frac{\partial \sum_{j=1}^{n} N_j p_j}{\partial x} + \frac{\partial}{\partial x} 2\eta \frac{\partial \sum_{j=1}^{n} N_j u_j}{\partial x} + \frac{\partial}{\partial y} \eta \left(\frac{\partial \sum_{j=1}^{n} N_j v_j}{\partial x} + \frac{\partial \sum_{j=1}^{n} N_j u_j}{\partial y} \right) \right]^2 + \\[2em] \left[-\frac{\partial \sum_{j=1}^{n} N_j p_j}{\partial y} + \frac{\partial}{\partial x} \eta \left(\frac{\partial \sum_{j=1}^{n} N_j u_j}{\partial y} + \frac{\partial \sum_{j=1}^{n} N_j v_j}{\partial x} \right) + \frac{\partial}{\partial y} 2\eta \frac{\partial \sum_{j=1}^{n} N_j v_j}{\partial y} \right]^2 + \\[2em] k \left[\left(\frac{\partial \sum_{j=1}^{n} N_j u_j}{\partial x} + \frac{\partial \sum_{j=1}^{n} N_j v_j}{\partial y} \right) \right]^2 \end{array} \right\} dx dy$$

$$(4.89)$$

where $N_j, j = 1,n$ are shape functions and n is the number of nodes per element. Minimization of functional (4.89) with respect to pressure and velocity components yields

$$
\begin{cases}
\dfrac{\partial I}{\partial u_j} = \displaystyle\int_{\Omega_e} 2\left[f_{\mathrm{res}}^u \left(\dfrac{\partial}{\partial x} 2\eta \dfrac{\partial N_j}{\partial x} + \dfrac{\partial}{\partial y}\eta \dfrac{\partial N_j}{\partial y} \right) + f_{\mathrm{res}}^v \left(\dfrac{\partial}{\partial x}\eta \dfrac{\partial N_j}{\partial y} \right) + kf_{\mathrm{res}}^c \dfrac{\partial N_j}{\partial x} \right] \mathrm{d}x\mathrm{d}y = 0 \\[4ex]
\dfrac{\partial I}{\partial v_j} = \displaystyle\int_{\Omega_e} 2\left[f_{\mathrm{res}}^u \left(\dfrac{\partial}{\partial y}\eta \dfrac{\partial N_j}{\partial x} \right) + f_{\mathrm{res}}^v \left(\dfrac{\partial}{\partial x}\eta \dfrac{\partial N_j}{\partial x} + \dfrac{\partial}{\partial y} 2\eta \dfrac{\partial N_j}{\partial y} \right) + kf_{\mathrm{res}}^c \dfrac{\partial N_j}{\partial y} \right] \mathrm{d}x\mathrm{d}y = 0 \\[4ex]
\dfrac{\partial I}{\partial p_j} = \displaystyle\int_{\Omega_e} 2\left(f_{\mathrm{res}}^u \dfrac{\partial N_j}{\partial x} + f_{\mathrm{res}}^v \dfrac{\partial N_j}{\partial y} \right) \mathrm{d}x\mathrm{d}y = 0
\end{cases}
$$

$$(4.90)$$

It is evident that application of Green's theorem cannot eliminate second-order derivatives of the shape functions in the set of working equations of the least-squares scheme. Therefore, direct application of these equations should, in general, be in conjunction with C^1 continuous Hermite elements (Petera and Nassehi, 1993; Petera and Pittman, 1994). However, various techniques are available that make the use of C^0 elements in these schemes possible. For example, Bell and Surana (1994) developed a method in which the flow model equations are cast into a set of auxiliary first-order differential equations. They used this approach to construct a least-squares scheme for non-Newtonian flow equations based on equal-order C^0 continuous, p-version hierarchical elements.

4.2   VARIATIONS OF VISCOSITY

Incorporation of viscosity variations in non-elastic generalized Newtonian flow models is based on using empirical rheological relationships such as the power law or Carreau equation, described in Chapter 1. In these relationships fluid viscosity is given as a function of shear rate and material parameters. Therefore in the application of finite element schemes to non-Newtonian flow, shear rate at the elemental level should be calculated and used to update the fluid viscosity. The shear rate is defined as the second invariant of the rate of deformation tensor as (Bird $et\ al.$, 1977)

$$\dot{\gamma} = \sqrt{\tfrac{1}{2}(\boldsymbol{D} : \boldsymbol{D})} \tag{4.91}$$

Equation (4.91) is written using the components of the rate of deformation tensor $\boldsymbol{D}$ as:

- in Cartesian (x, y) coordinate system

$$\dot{\gamma} = \left[2\left(\frac{\partial v_x}{\partial x}\right)^2 + 2\left(\frac{\partial v_y}{\partial y}\right)^2 + \left(\frac{\partial v_x}{\partial y} + \frac{\partial v_y}{\partial x}\right)^2 \right]^{\frac{1}{2}} \tag{4.92}$$

- in polar (r, θ) coordinate system

$$\dot{\gamma} = \left[2\left(\frac{\partial v_r}{\partial r}\right)^2 + \frac{2}{r^2}\left(\frac{\partial v_\theta}{\partial \theta} + v_r\right)^2 + \left(\frac{\partial v_\theta}{\partial r} - \frac{v_\theta}{r} + \frac{1}{r}\frac{\partial v_r}{\partial \theta}\right)^2 \right]^{\frac{1}{2}} \tag{4.93}$$

- in axisymmetric (r, z) coordinate system

$$\dot{\gamma} = \left[2\left(\frac{\partial v_r}{\partial r}\right)^2 + 2\left(\frac{v_r}{r}\right)^2 + 2\left(\frac{\partial v_z}{\partial z}\right)^2 + \left(\frac{\partial v_r}{\partial z} + \frac{\partial v_z}{\partial r}\right)^2 \right]^{\frac{1}{2}} \tag{4.94}$$

4.3 MODELLING OF STEADY-STATE VISCOMETRIC FLOW – WORKING EQUATIONS OF THE CONTINUOUS PENALTY SCHEME IN CARTESIAN COORDINATE SYSTEMS

The (CEF) model (see Chapter 1) provides a simple means for obtaining useful results for steady-state viscometric flow of polymeric fluids (Tanner, 1985). In this approach the extra stress in the equation of motion is replaced by explicit relationships in terms of rate of strain components. For example, assuming a zero second normal stress difference for very slow flow regimes such relationships are written as (Mitsoulis *et al.*, 1985)

(for flow dominantly in the x direction)

$$\begin{cases} \tau_{xx} = 2\eta_0 \frac{\partial u}{\partial x} + \Psi_1 \left(\frac{\partial u}{\partial y} + \frac{\partial v}{\partial x}\right)^2 = \eta_0 \dot{\gamma}_{xx} + \Psi_1 \dot{\gamma}_{xy}^2 \\ \tau_{xy} = \eta \left(\frac{\partial u}{\partial y} + \frac{\partial v}{\partial x}\right) = \dot{\gamma}_{xy} \\ \tau_{yy} = 2\eta_0 \frac{\partial v}{\partial y} = \eta_0 \dot{\gamma}_{yy} \end{cases} \tag{4.95}$$

where η_0 is the fluid consistency coefficient, η is shear-dependent viscosity and Ψ_1 is a material parameter representing the first normal stress difference. This parameter can be defined empirically as

$$\Psi_1 = A\eta^b |\dot{\gamma}_{xy}|^{b-2} \tag{4.96}$$

where A and b and are characteristic material constants. After substitution of the extra stress term in the equation of motion through the relationships given above and discretization of the resulting equations according to the previously described continuous penalty method the working equations of this scheme are obtained as

$$\begin{bmatrix} A_{ij}^{11} & A_{ij}^{12} \\ A_{ij}^{21} & A_{ij}^{22} \end{bmatrix} \begin{Bmatrix} u_j \\ v_j \end{Bmatrix} = \begin{Bmatrix} B_j^1 \\ B_j^2 \end{Bmatrix} \tag{4.97}$$

where

$$A_{ij}^{11} = \int_{\Omega_e} \left[(\lambda + 2\eta_0) \frac{\partial N_i}{\partial x} \frac{\partial N_j}{\partial x} + \eta \frac{\partial N_i}{\partial y} \frac{\partial N_j}{\partial y} + \Psi_1 \left(\frac{\partial N_i}{\partial x} \frac{\partial N_j}{\partial y} \right) \left(\frac{\partial \bar{u}^e}{\partial y} + \frac{\partial \bar{v}^e}{\partial x} \right) \right] \mathrm{d}x\mathrm{d}y$$
$$\tag{4.98}$$

$$A_{ij}^{12} = \int_{\Omega_e} \left[\lambda \frac{\partial N_i}{\partial x} \frac{\partial N_j}{\partial y} + \eta \frac{\partial N_i}{\partial y} \frac{\partial N_j}{\partial x} + \Psi_1 \left(\frac{\partial N_i}{\partial x} \frac{\partial N_j}{\partial x} \right) \left(\frac{\partial \bar{u}^e}{\partial y} + \frac{\partial \bar{v}^e}{\partial x} \right) \right] \mathrm{d}x\mathrm{d}y \tag{4.99}$$

$$A_{ij}^{21} = \int_{\Omega_e} \left(\lambda \frac{\partial N_i}{\partial y} \frac{\partial N_j}{\partial x} + \eta \frac{\partial N_i}{\partial x} \frac{\partial N_j}{\partial y} \right) \mathrm{d}x\mathrm{d}y \tag{4.100}$$

$$A_{ij}^{22} = \int_{\Omega_e} \left[(\lambda + 2\eta_0) \frac{\partial N_i}{\partial y} \frac{\partial N_j}{\partial y} + \eta \frac{\partial N_i}{\partial x} \frac{\partial N_j}{\partial x} \right] \mathrm{d}x\mathrm{d}y \tag{4.101}$$

$$B_j^1 = \int_{\Gamma_e} N_i \left\{ \left[-p^e + 2\eta_0 \frac{\partial u^e}{\partial x} + \Psi_1 \left(\frac{\partial \bar{u}^e}{\partial y} + \frac{\partial \bar{v}^e}{\partial x} \right)^2 \right] n_x + \eta \left(\frac{\partial u^e}{\partial y} + \frac{\partial v^e}{\partial x} \right) n_y \right\} \mathrm{d}\Gamma_e$$
$$\tag{4.102}$$

$$B_j^2 = \int_{\Gamma_e} N_i \left[\eta \left(\frac{\partial u^e}{\partial y} + \frac{\partial v^e}{\partial x} \right) n_x + \left(-p^e + 2\eta_0 \frac{\partial v^e}{\partial y} \right) n_y \right] \mathrm{d}\Gamma_e \tag{4.103}$$

where an over bar indicates a value found at the last iteration step.

4.4   MODELLING OF THERMAL ENERGY BALANCE

The majority of polymer flow processes involve significant heat dissipation and should be regarded as non-isothermal regimes. Therefore in the finite element modelling of polymeric flow, in conjunction with the equations of continuity and motion, an appropriate heat balance equation should be solved. Due to the temperature dependency of the fluid viscosity, coupling of the flow and energy equations introduces an additional form of non-linearity into the non-Newtonian polymer flow models. The degree of this non-linearity and inter-dependency between the equations of motion and energy is determined by the magnitudes of the Nahme–Griffith number and the Pearson number in the flow system. These dimensionless numbers are defined as

$$Na = \frac{\mu u^2 \xi}{k} \tag{4.104}$$

$$Pn = (\Delta T)_{\text{op}} \xi \tag{4.105}$$

where μ and u are characteristic viscosity and velocity in the flow domain, ξ is the temperature dependency coefficient of viscosity (parameter b in Equation (1.14)), k is fluid conductivity and $(\Delta T)_{\text{op}}$ is the operationally imposed temperature difference across the boundaries. Physically, the Nahme–Griffith number gives a measure of the ratio of viscous heat dissipation to the heat required to significantly alter the fluid viscosity while the Pearson number provides a measure for the influence on viscosity of the imposed temperature difference through the flow domain boundaries. Analysis of these factors has shown that, in general, the temperature dependency of viscosity in polymeric flow processes cannot be neglected (Pearson, 1979; Pittman and Nakazawa, 1984). In practice, however, coupling of the flow and energy equations can be based on an iterative algorithm. Details of such an algorithm are described in Chapter 7.

4.4.1 Working equations of the streamline upwind (SU) scheme for the steady-state energy equation in Cartesian, polar and axisymmetric coordinate systems

Following the procedure described in in Chapter 3, Section 3 the streamlined-upwind 'weighted residual' statement of the energy equation is formulated as

$$\int_{\Omega_e} \left[\rho c N_i^* \left(u^e \frac{\partial \sum_{j=1}^n N_j T_j}{\partial x} + v^e \frac{\partial \sum_{j=1}^n N_j T_j}{\partial y} \right) - N_i \left(\frac{\partial}{\partial x} k \frac{\partial \sum_{j=1}^n N_j T_j}{\partial x} + \frac{\partial}{\partial y} k \frac{\partial \sum_{j=1}^n N_j T_j}{\partial y} + \eta \dot{\gamma}^2 \right) \right] \mathrm{d}x\mathrm{d}y = 0 \tag{4.106}$$

where N_i is the weight function (identical to N_j elemental shape functions) and N_i^* is the upwinded weight functions defined according to the streamline upwind formula for a rectangular element as

$$N_i^* = N_i + \varepsilon \frac{|h_x u^e + h_y v^e|}{2|V|^2} \left(u^e \frac{\partial N_i}{\partial x} + v^e \frac{\partial N_i}{\partial y} \right) \tag{4.107}$$

where $0 < \varepsilon \le 1$ is the upwinding constant and characteristic element dimensions are defined as

$$\begin{cases} h_x = 2\dfrac{\partial x^e}{\partial \xi} + 2\dfrac{\partial x^e}{\partial \eta} \\[2mm] h_y = 2\dfrac{\partial y^e}{\partial \xi} + 2\dfrac{\partial y^e}{\partial \eta} \end{cases} \tag{4.108}$$

where ξ and η and are the elemental coordinates. As mentioned before the definition of the upwinding parameter (i.e. coefficient multiplied by the derivatives of the shape functions in Equation (4.107) is based on a heuristic analogy with one-dimensional analysis and other forms are also used (e.g. see Equation (3.33)).

After the application of Green's theorem to the second-order derivatives of temperature in Equation (4.106) the working equation of this scheme is obtained as

$$[A_{ij}]\{T_j\} = \{B_j\} \tag{4.109}$$

where in a Cartesian coordinate system

$$A_{ij} = \int_{\Omega_e} \left[\rho\, c\, N_i^* \left(u^e \frac{\partial N_j}{\partial x} + v^e \frac{\partial N_j}{\partial y} \right) + k \left(\frac{\partial N_i}{\partial x} \frac{\partial N_j}{\partial x} + \frac{\partial N_i}{\partial y} \frac{\partial N_j}{\partial y} \right) \right] dxdy \tag{4.110}$$

$$B_j = \int_{\Omega_e} N_i [\eta(\dot{\gamma}^2)^e] dxdy + \int_{\Gamma_e} N_i \left(k \frac{\partial T^e}{\partial x} n_x + k \frac{\partial T^e}{\partial y} n_y \right) d\Gamma_e \tag{4.111}$$

In a polar coordinate system the steady-state energy equation is written as

$$\rho\, c \left(v_r \frac{\partial T}{\partial r} + \frac{v_\theta}{r} \frac{\partial T}{\partial \theta} \right) = \frac{1}{r} \frac{\partial}{\partial r} \left(rk \frac{\partial T}{\partial r} \right) + \frac{1}{r^2} \frac{\partial}{\partial \theta} \left(k \frac{\partial T}{\partial \theta} \right) + \eta \dot{\gamma}^2 \tag{4.112}$$

Using a similar discretization the terms of the stiffness and load matrices corresponding to Equation (4.112) are written in this scheme as

$$A_{ij} = \int_{\Omega_e} \left[\rho\, c\, N_i^* \left(v_r^e \frac{\partial N_j}{\partial r} + \frac{v_\theta^e}{r} \frac{\partial N_j}{\partial \theta} \right) + k \left(\frac{\partial N_i}{\partial r} \frac{\partial N_j}{\partial r} + \frac{1}{r^2} \frac{\partial N_i}{\partial \theta} \frac{\partial N_j}{\partial \theta} \right) \right] rdrd\theta \tag{4.113}$$

$$B_j = \int_{\Omega_e} N_i[\eta(\dot{\gamma}^2)^e] r dr d\theta + \int_{\Gamma_e} N_i \left(k \frac{\partial T^e}{\partial r} n_r + \frac{k}{r^2} \frac{\partial T^e}{\partial \theta} n_\theta \right) r d\Gamma_e \qquad (4.114)$$

Similarly in an axisymmetric coordinate system the terms of stiffness and load matrices corresponding to the governing energy equation written as

$$\rho c \left(v_r \frac{\partial T}{\partial r} + v_z \frac{\partial T}{\partial z} \right) = \frac{1}{r} \frac{\partial}{\partial r} \left(rk \frac{\partial T}{\partial r} \right) + \frac{\partial}{\partial z} \left(k \frac{\partial T}{\partial z} \right) + \eta \dot{\gamma}^2 \qquad (4.115)$$

are derived as

$$A_{ij} = \int_{\Omega_e} \left[\rho c N_i^* \left(v_r^e \frac{\partial N_j}{\partial r} + v_z^e \frac{\partial N_j}{\partial z} \right) + k \left(\frac{\partial N_i}{\partial r} \frac{\partial N_j}{\partial r} + \frac{\partial N_i}{\partial z} \frac{\partial N_j}{\partial z} \right) \right] r dr dz \quad (4.116)$$

$$B_j = \int_{\Omega_e} N_i[\eta(\dot{\gamma}^2)^e] r dr dz + \int_{\Gamma_e} k N_i \left(\frac{\partial T^e}{\partial r} n_r + \frac{\partial T^e}{\partial z} n_z \right) r d\Gamma_e \qquad (4.117)$$

4.4.2 Least-squares and streamline upwind Petrov–Galerkin (SUPG) schemes

The inconsistent streamline upwind scheme described in the last section is formulated in an ad hoc manner and does not correspond to a weighted residual statement in a strict sense. In this section we consider the development of weighted residual schemes for the finite element solution of the energy equation. Using vector notation for simplicity the energy equation is written as

$$\rho c \left(\frac{\partial T}{\partial t} + v.\nabla T \right) - \nabla.k\nabla T - \dot{q} = 0 \qquad (4.118)$$

After application of the θ time-stepping method (see Chapter 2, Section 2.5) and following the procedure outlined in Chapter 2, Section 2.4, a functional representing the sum of the squares of the approximation error generated by the finite element discretization of Equation (4.118) is formulated as

$$I_{n+1} = \int_{\Omega_e} \left[\rho c \frac{\sum_{j=1}^{p} N_j T_j^{n+1} - \sum_{j=1}^{p} N_j T_j^n}{\Delta t_n} + \rho c v.\nabla \left(\sum_{j=1}^{p} N_j T_j^{n+\theta} \right) - \right.$$
$$\left. \nabla.k\nabla \left(\sum_{j=1}^{p} N_j T_j^{n+\theta} \right) - \dot{q} \right]^2 d\Omega_e = \int_{\Omega_e} [f_{res}^T]^2 d\Omega_e \qquad (4.119)$$

where n is the time level, $N_j, j = 1, p$ are the shape functions and ρ, c, v, k and $\dot{q}$ are all given in Ω_e. Minimization of functional (4.119) with respect to nodal temperatures gives

$$\frac{\partial I}{\partial T_j} = 2 \int_{\Omega_e} \left(\rho c \frac{N_j}{\Delta t_n} + \rho c \theta v . \nabla N_j - \theta \nabla . k \nabla N_j \right) [f_{\text{res}}^T] d\Omega_e = 0 \qquad (4.120)$$

Assuming constant density and specific heat (i.e. ρc = constant) Equation (4.120) is written as

$$\int_{\Omega_e} (N_j + \theta \Delta t_n v . \nabla N_j - \theta \Delta t_n \nabla . \Theta \nabla N_j) [f_{\text{res}}^T] d\Omega_e = 0 \qquad (4.121)$$

where $\Theta = k/\rho c$ is the thermal diffusivity. Equation (4.121) represents a weighted residual statement where the weighting function is given as

$$W = N_j + \theta \Delta t_n v . \nabla N_j - \theta \Delta t_n \nabla . \Theta \nabla N_j \qquad (4.122)$$

The following options can now be considered:

- If the second and third terms in the weight function are neglected the standard Galerkin scheme will be obtained.

- If only the third term in the weight function is neglected a first-order Petrov–Galerkin scheme corresponding to the SUPG method will be obtained. Inconsistent upwinding will be a special case in which the second term in the weight function is only retained for the weighting of the convection terms.

- Retaining all of the terms in the weight function a least-squares scheme corresponding to a second-order Petrov–Galerkin formulation will be obtained.

- In steady-state problems $\theta \Delta t_n = 1$ and the time-dependent term in the residual is eliminated. The steady-state scheme will hence be equivalent to the combination of Galerkin and least-squares methods.

4.5 MODELLING OF TRANSIENT STOKES FLOW OF GENERALIZED NEWTONIAN AND NON-NEWTONIAN FLUIDS

In the absence of body force the equations of continuity and motion representing transient Stokes flow of a generalized Newtonian fluid in a two-dimensional Cartesian system are written, on the basis of Equations (1.1) and (1.4), as

$$\frac{\partial u}{\partial x} + \frac{\partial v}{\partial y} = 0 \qquad \text{(continuity)} \tag{4.123}$$

where u and v are the x and y components of velocity, respectively, and

$$\begin{cases} \rho \dfrac{\partial u}{\partial t} = -\dfrac{\partial p}{\partial x} + \dfrac{\partial}{\partial x}\left(2\eta \dfrac{\partial u}{\partial x}\right) + \dfrac{\partial}{\partial x}\left[\eta\left(\dfrac{\partial v}{\partial x} + \dfrac{\partial u}{\partial y}\right)\right] \\[4mm] \rho \dfrac{\partial v}{\partial t} = -\dfrac{\partial p}{\partial y} + \dfrac{\partial}{\partial x}\left[\eta\left(\dfrac{\partial u}{\partial y} + \dfrac{\partial v}{\partial x}\right)\right] + \dfrac{\partial}{\partial y}\left(2\eta \dfrac{\partial v}{\partial y}\right) \end{cases} \qquad \text{(motion)} \tag{4.124}$$

where p is pressure. Following the partial discretization technique, described in Chapter 2, Section 2.5, various finite element schemes for this problem can be developed. As an example, we consider derivation of the working equations of the continuous penalty scheme in conjunction with the θ time-stepping procedure. After substitution of pressure in equation set (4.124) via the penalty relationship, and spatial discretization of the resulting system, in a manner similar to the procedure described in Section 1.7, we obtain

$$\begin{bmatrix} C_{ij}^{11} & 0 \\ 0 & C_{ij}^{22} \end{bmatrix} \begin{Bmatrix} \dot{u}_j \\ \dot{v}_j \end{Bmatrix} + \begin{bmatrix} A_{ij}^{11} & A_{ij}^{12} \\ A_{ij}^{21} & A_{ij}^{22} \end{bmatrix} \begin{Bmatrix} u_j \\ v_j \end{Bmatrix} = \begin{Bmatrix} B_j^1 \\ B_j^2 \end{Bmatrix} \tag{4.125}$$

where

$$C_{ij}^{11} = C_{ij}^{22} = \int_{\Omega_e} \rho N_i N_j \mathrm{d}x\mathrm{d}y \tag{4.126}$$

The remaining terms in equation set (4.125) are identical to their counterparts derived for the steady-state case (given as Equations (4.55) to (4.60)). By application of the θ time-stepping method, described in Chapter 2, Section 2.5, to the set of first-order ordinary differential equations (4.125) the working equations of the solution scheme are obtained. The general form of these equations will be identical to Equation (2.111) in Chapter 2.

The described continuous penaltyθ time-stepping scheme may yield unstable results in some problems. Therefore we consider an alternative scheme which provides better numerical stability under a wide range of conditions. This scheme is based on the U–V–P method for the slightly compressible continuity equation, described in Chapter 3, Section 1.2, in conjunction with the Taylor–Galerkin time-stepping (see Chapter 2, Section 2.5). The governing equations used in this scheme are as follows

$$\frac{1}{\rho c^2}\frac{\partial p}{\partial t} + \frac{\partial u}{\partial x} + \frac{\partial v}{\partial y} = 0 \tag{4.127}$$

(continuity equation, based on Equation (3.4))

$$\begin{cases} \rho \dfrac{\partial u}{\partial t} = -\dfrac{\partial p}{\partial x} + \dfrac{\partial \tau_{xx}}{\partial x} + \dfrac{\partial \tau_{xy}}{\partial y} \\[3mm] \rho \dfrac{\partial v}{\partial t} = -\dfrac{\partial p}{\partial x} + \dfrac{\partial \tau_{yx}}{\partial x} + \dfrac{\partial \tau_{yy}}{\partial y} \end{cases} \tag{4.128}$$

(equation of motion for Stokes flow based on Equation (1.4))

As explained in Chapter 3, it is possible to use equal order interpolation models for the spatial discretization of velocity and pressure in a U–V–P scheme based on Equations (4.127) and (4.128) without violating the BB stability condition.

To develop the scheme we start with the normalization of the governing equations by letting

$$\begin{cases} U = u, V = v \\[3mm] P = \dfrac{p}{\rho} \\[3mm] T_{xx} = \dfrac{\tau_{xx}}{\rho}, T_{xy} = T_{yx} = \dfrac{\tau_{xy}}{\rho}, T_{yy} = \dfrac{\tau_{yy}}{\rho} \end{cases} \tag{4.129}$$

Thus we have

$$\frac{1}{c^2} \frac{\partial P}{\partial t} + \frac{\partial U}{\partial x} + \frac{\partial V}{\partial y} = 0 \tag{4.130}$$

and

$$\begin{cases} \dfrac{\partial U}{\partial t} = -\dfrac{\partial P}{\partial x} + \dfrac{\partial T_{xx}}{\partial x} + \dfrac{\partial T_{xy}}{\partial y} \\[3mm] \dfrac{\partial V}{\partial t} = -\dfrac{\partial P}{\partial x} + \dfrac{\partial T_{yx}}{\partial x} + \dfrac{\partial T_{yy}}{\partial y} \end{cases} \tag{4.131}$$

Following the procedure described in Chatper 2, Section 2.5 the Taylor series expansion of the field unknowns at a time level equal to $n + \alpha\Delta t$, where $0 \le \alpha \le 1$, are obtained as

$$\frac{\Delta P}{\Delta t} = \frac{P|_{n+1} - P|_n}{\Delta t} = \frac{\partial P}{\partial t}\bigg|_{n+\alpha\Delta t} + \tfrac{1}{2}\alpha\Delta t \frac{\partial^2 P}{\partial t^2}\bigg|_{n+\alpha\Delta t} \tag{4.132}$$

$$\frac{\Delta U}{\Delta t} = \frac{U|_{n+1} - U|_n}{\Delta t} = \frac{\partial U}{\partial t}\bigg|_{n+\alpha\Delta t} + \tfrac{1}{2}\alpha\Delta t \frac{\partial^2 U}{\partial t^2}\bigg|_{n+\alpha\Delta t} \tag{4.133}$$

$$\frac{\Delta V}{\Delta t} = \frac{V|_{n+1} - V|_n}{\Delta t} = \frac{\partial V}{\partial t}\bigg|_{n+\alpha\Delta t} + \tfrac{1}{2}\alpha\Delta t \frac{\partial^2 V}{\partial t^2}\bigg|_{n+\alpha\Delta t} \tag{4.134}$$

The selection of a time increment dependent on parameter α (i.e. carrying out Taylor series expansion at a level between successive time steps of n and $n+1$) enhances the flexibility of the temporal discretizations by allowing the introduction of various amounts of smoothing in different problems. The first-order time derivatives are found from the governing equations as

$$\frac{\partial P}{\partial t}\bigg|_{n+\alpha\Delta t} = -c^2\left(\frac{\partial U}{\partial x} + \frac{\partial V}{\partial y}\right)\bigg|_{n+\alpha\Delta t} \tag{4.135}$$

$$\frac{\partial U}{\partial t}\bigg|_{n+\alpha\Delta t} = -\frac{\partial P}{\partial x}\bigg|_{n+\alpha\Delta t} + \left(\frac{\partial T_{xx}}{\partial x} + \frac{\partial T_{xy}}{\partial y}\right)\bigg|_{n+\alpha\Delta t} \tag{4.136}$$

$$\frac{\partial V}{\partial t}\bigg|_{n+\alpha\Delta t} = -\frac{\partial P}{\partial y}\bigg|_{n+\alpha\Delta t} + \left(\frac{\partial T_{yx}}{\partial x} + \frac{\partial T_{yy}}{\partial y}\right)\bigg|_{n+\alpha\Delta t} \tag{4.137}$$

The second-order derivatives of the variables are now found as

$$\frac{\partial^2 U}{\partial t^2}\bigg|_{n+\alpha\Delta t} = \frac{\partial}{\partial t}\left(\frac{\partial U}{\partial t}\right)\bigg|_{n+\alpha\Delta t} = -\frac{\partial}{\partial t}\left(\frac{\partial P}{\partial x}\right)\bigg|_{n+\alpha\Delta t} + \frac{\partial}{\partial t}\left(\frac{\partial T_{xx}}{\partial x} + \frac{\partial T_{xy}}{\partial y}\right)\bigg|_{n+\alpha\Delta t} =$$

$$-\frac{\partial}{\partial x}\left(\frac{\partial P}{\partial t}\right)\bigg|_{n+\alpha\Delta t} + \frac{\partial}{\partial t}\left(\frac{\partial T_{xx}}{\partial x} + \frac{\partial T_{xy}}{\partial y}\right)\bigg|_{n+\alpha\Delta t} =$$

$$c^2\frac{\partial}{\partial x}\left(\frac{\partial U}{\partial x} + \frac{\partial V}{\partial y}\right)\bigg|_{n+\alpha\Delta t} + \frac{\partial}{\partial t}\left(\frac{\partial T_{xx}}{\partial x} + \frac{\partial T_{xy}}{\partial y}\right)\bigg|_{n+\alpha\Delta t} \tag{4.138}$$

The extra stress is proportional to the derivatives of velocity components and consequently the order of velocity derivatives in terms arising from

$$\frac{\partial}{\partial t}\left(\frac{\partial T_{xx}}{\partial x} + \frac{\partial T_{xy}}{\partial y}\right)\bigg|_{n+\alpha\Delta t}$$

will be higher than the term

$$c^2\frac{\partial}{\partial x}\left(\frac{\partial U}{\partial x} + \frac{\partial V}{\partial y}\right)\bigg|_{n+\alpha\Delta t}$$

hence the stress terms in Equation (4.138) can be ignored to obtain

$$\frac{\partial^2 U}{\partial t^2}\Big|_{n+\alpha\Delta t} = c^2 \frac{\partial}{\partial x}\left(\frac{\partial U}{\partial x} + \frac{\partial V}{\partial y}\right)\Big|_{n+\alpha\Delta t} \tag{4.139}$$

Similarly

$$\frac{\partial^2 V}{\partial t^2}\Big|_{n+\alpha\Delta t} = c^2 \frac{\partial}{\partial y}\left(\frac{\partial U}{\partial x} + \frac{\partial V}{\partial y}\right)\Big|_{n+\alpha\Delta t} \tag{4.140}$$

Note that in polar and axisymmetric coordinate systems the stress term will include some lower-order terms that should be included in the formulations.

Using a similar procedure the second order time derivative of pressure is found as

$$\frac{\partial^2 P}{\partial t^2}\Big|_{n+\alpha\Delta t} = \frac{\partial}{\partial t}\left(\frac{\partial P}{\partial t}\right)\Big|_{n+\alpha\Delta t} = \frac{\partial}{\partial t}\left[-c^2\left(\frac{\partial U}{\partial x} + \frac{\partial V}{\partial y}\right)\right]\Big|_{n+\alpha\Delta t} =$$

$$-c^2\left[\frac{\partial}{\partial x}\left(\frac{\partial U}{\partial t}\right) + \frac{\partial}{\partial y}\left(\frac{\partial V}{\partial t}\right)\right]\Big|_{n+\alpha\Delta t} = c^2\frac{\partial}{\partial x}\left[\frac{\partial P}{\partial x} - \left(\frac{\partial T_{xx}}{\partial x} + \frac{\partial T_{xy}}{\partial y}\right)\right]\Big|_{n+\alpha\Delta t} +$$

$$c^2\frac{\partial}{\partial y}\left[\frac{\partial P}{\partial y} - \left(\frac{\partial T_{yx}}{\partial x} + \frac{\partial T_{yy}}{\partial y}\right)\right]\Big|_{n+\alpha\Delta t} \tag{4.141}$$

After substitution of the first- and second-order time derivatives of the unknowns in Equations (4.132) to (4.134) from Equations (4.139) to (4.141) and spatial discretization of the resulting equations in the usual manner the working equations of the scheme are derived. In these equations, functions given at time level $n+\alpha\Delta t$ can be interpolated as

$$A|_{n+\alpha\Delta t} = \alpha A|_{n+1} + (1 - \alpha)A|_n \tag{4.142}$$

In generalized Newtonian fluids, before derivation of the final set of the working equations, the extra stress in the expanded equations should be replaced using the components of the rate of strain tensor (note that the viscosity should also be normalized as $\bar{\eta} = \eta/\rho$). In contrast, in the modelling of viscoelastic fluids, stress components are found at a separate step through the solution of a constitutive equation. This allows the development of a robust Taylor–Galerkin/U–V–P scheme on the basis of the described procedure in which the stress components are all found at time level n. The final working equation of this scheme can be expressed as

$$
\begin{bmatrix}
M_{ij}^{11} & M_{ij}^{12} & M_{ij}^{13} \\
M_{ij}^{21} & M_{ij}^{22} & M_{ij}^{23} \\
M_{ij}^{31} & M_{ij}^{32} & M_{ij}^{33}
\end{bmatrix}^{n+1}
\left\{
\begin{array}{c}
U_j \\
V_j \\
P_j
\end{array}
\right\}^{n+1}
=
$$

$$
\begin{bmatrix}
K_{ij}^{11} & K_{ij}^{12} & K_{ij}^{13} \\
K_{ij}^{21} & K_{ij}^{22} & K_{ij}^{23} \\
K_{ij}^{31} & K_{ij}^{32} & K_{ij}^{33}
\end{bmatrix}^{n}
\left\{
\begin{array}{c}
U_j \\
V_j \\
P_j
\end{array}
\right\}^{n}
+
\left\{
\begin{array}{c}
C_j^1 \\
C_j^2 \\
C_j^3
\end{array}
\right\}^{n}
+
\left\{
\begin{array}{c}
B_j^1 \\
B_j^2 \\
B_j^3
\end{array}
\right\}^{n+\alpha \Delta t}
\tag{4.143}
$$

where

$$
M_{ij}^{11} = \int_{\Omega_e} \left[N_i N_j + \tfrac{1}{2}\alpha(\Delta t)^2 c^2 \frac{\partial N_i}{\partial x}\frac{\partial N_j}{\partial x} \right] dxdy \tag{4.144}
$$

$$
M_{ij}^{12} = \int_{\Omega_e} \tfrac{1}{2}\alpha(\Delta t)^2 c^2 \frac{\partial N_i}{\partial x}\frac{\partial N_j}{\partial y}\, dxdy \tag{4.145}
$$

$$
M_{ij}^{13} = \int_{\Omega_e} -\alpha \Delta t \frac{\partial N_i}{\partial x} N_j\, dxdy \tag{4.146}
$$

$$
M_{ij}^{21} = \int_{\Omega_e} \tfrac{1}{2}\alpha(\Delta t)^2 c^2 \frac{\partial N_i}{\partial y}\frac{\partial N_j}{\partial x}\, dxdy \tag{4.147}
$$

$$
M_{ij}^{22} = \int_{\Omega_e} \left[N_i N_j + \tfrac{1}{2}\alpha(\Delta t)^2 c^2 \frac{\partial N_i}{\partial y}\frac{\partial N_j}{\partial y} \right] dxdy \tag{4.148}
$$

$$
M_{ij}^{23} = \int_{\Omega_e} -\alpha \Delta t \frac{\partial N_i}{\partial y} N_j\, dxdy \tag{4.149}
$$

$$
M_{ij}^{31} = \int_{\Omega_e} -\alpha \Delta t N_i \frac{\partial N_j}{\partial x}\, dxdy \tag{4.150}
$$

$$
M_{ij}^{32} = \int_{\Omega_e} -\alpha \Delta t N_i \frac{\partial N_j}{\partial y}\, dxdy \tag{4.151}
$$

$$
M_{ij}^{33} = \int_{\Omega_e} \left[\frac{-1}{c^2} N_i N_j - \tfrac{1}{2}\alpha(\Delta t)^2 \left(\frac{\partial N_i}{\partial x}\frac{\partial N_j}{\partial x} + \frac{\partial N_i}{\partial y}\frac{\partial N_j}{\partial y} \right) \right] dxdy \tag{4.152}
$$

$$
K_{ij}^{11} = \int_{\Omega_e} \left[N_i N_j - \tfrac{1}{2}(1-\alpha)(\Delta t)^2 c^2 \frac{\partial N_i}{\partial x}\frac{\partial N_j}{\partial x} \right] dxdy \tag{4.153}
$$

$$K_{ij}^{12} = \int_{\Omega_e} \left[-\tfrac{1}{2}(1-\alpha)(\Delta t)^2 c^2 \frac{\partial N_i}{\partial x} \frac{\partial N_j}{\partial y} \right] dx dy \tag{4.154}$$

$$K_{ij}^{13} = \int_{\Omega_e} (1-\alpha)\Delta t \frac{\partial N_i}{\partial x} N_j \, dx dy \tag{4.155}$$

$$K_{ij}^{21} = \int_{\Omega_e} \left[-\tfrac{1}{2}(1-\alpha)(\Delta t)^2 c^2 \frac{\partial N_i}{\partial y} \frac{\partial N_j}{\partial x} \right] dx dy \tag{4.156}$$

$$K_{ij}^{22} = \int_{\Omega_e} \left[N_i N_j - \tfrac{1}{2}(1-\alpha)(\Delta t)^2 c^2 \frac{\partial N_i}{\partial y} \frac{\partial N_j}{\partial y} \right] dx dy \tag{4.157}$$

$$K_{ij}^{23} = \int_{\Omega_e} (1-\alpha)\Delta t \frac{\partial N_i}{\partial y} N_j \, dx dy \tag{4.158}$$

$$K_{ij}^{31} = \int_{\Omega_e} (1-\alpha)\Delta t N_i \frac{\partial N_j}{\partial x} \, dx dy \tag{4.159}$$

$$K_{ij}^{32} = \int_{\Omega_e} (1-\alpha)\Delta t N_i \frac{\partial N_j}{\partial y} \, dx dy \tag{4.160}$$

$$K_{ij}^{33} = \int_{\Omega_e} \left[\frac{-1}{c^2} N_i N_j + \tfrac{1}{2}(1-\alpha)(\Delta t)^2 \left(\frac{\partial N_i}{\partial x} \frac{\partial N_j}{\partial x} + \frac{\partial N_i}{\partial y} \frac{\partial N_j}{\partial y} \right) \right] dx dy \tag{4.161}$$

$$C_j^1 = \int_{\Omega_e} -\left(T_{xx}^e \frac{\partial N_i}{\partial x} + T_{xy}^e \frac{\partial N_i}{\partial y} \right) \Delta t \, dx dy \tag{4.162}$$

$$C_j^2 = \int_{\Omega_e} -\left(T_{yx}^e \frac{\partial N_i}{\partial x} + T_{yy}^e \frac{\partial N_i}{\partial y} \right) \Delta t \, dx dy \tag{4.163}$$

$$C_j^3 = \int_{\Omega_e} -\tfrac{1}{2}\alpha \left[\left(\frac{\partial T_{xx}^e}{\partial x} + \frac{\partial T_{xy}^e}{\partial y} \right) \frac{\partial N_i}{\partial x} + \left(\frac{\partial T_{yx}^e}{\partial x} + \frac{\partial T_{yy}^e}{\partial y} \right) \frac{\partial N_i}{\partial y} \right] (\Delta t)^2 \, dx dy \tag{4.164}$$

The last term in the right-hand side of Equation (4.143) represents boundary line integrals. These result from the application of Green's theorem to

second-order derivatives of velocity
first-order derivatives of pressure   } (in equations corresponding to U and V)
first-order derivatives of stress

and

second-order derivatives of pressure $\Big\}$ (in equation corresponding to P)
second-order derivatives of stress

Therefore

$$B_j^1 = \int_{\Gamma_e} N_i \left[\tfrac{1}{2}\alpha(\Delta t)^2 c^2 \frac{\partial U^e}{\partial x} + \tfrac{1}{2}\alpha(\Delta t)^2 c^2 \frac{\partial V^e}{\partial y} - \Delta t P^e \right] n_x \bigg|_{n+\alpha\Delta t} \mathrm{d}\Gamma_e +$$

$$\int_{\Gamma_e} \Delta t N_i (T_{xx}n_x + T_{xy}n_y)|_n \, \mathrm{d}\Gamma_e \qquad (4.165)$$

$$B_j^2 = \int_{\Gamma_e} N_i \left[\tfrac{1}{2}\alpha(\Delta t)^2 c^2 \frac{\partial U^e}{\partial x} + \tfrac{1}{2}\alpha(\Delta t)^2 c^2 \frac{\partial V^e}{\partial y} - \Delta t P^e \right] n_y \bigg|_{n+\alpha\Delta t} \mathrm{d}\Gamma_e +$$

$$\int_{\Gamma_e} \Delta t N_i (T_{yx}n_x + T_{yy}n_y)|_n \, \mathrm{d}\Gamma_e \qquad (4.166)$$

$$B_j^3 = \int_{\Gamma_e} -\tfrac{1}{2}\alpha\left(\Delta t \right)^2 N_i \left(\frac{\partial P^e}{\partial x} n_x + \frac{\partial P^e}{\partial y} n_y \right) \bigg|_{n+\alpha\Delta t} \mathrm{d}\Gamma_e +$$

$$\int_{\Gamma_e} +\tfrac{1}{2}\alpha(\Delta t)^2 N_i \left[\left(\frac{\partial T_{xx}}{\partial x} + \frac{\partial T_{xy}}{\partial y} \right) n_x + \left(\frac{\partial T_{yx}}{\partial x} + \frac{\partial T_{yy}}{\partial y} \right) n_y \right] \bigg|_n \mathrm{d}\Gamma_e \qquad (4.167)$$

The described scheme can also be incorporated into iterative algorithms and used to solve steady-state flow problems (Zienkiewicz and Wu, 1991).

REFERENCES

Bell, B. C. and Surana, K. S., 1994. p-version least squares finite element formulations for two-dimensional, incompressible, non-Newtonian isothermal and non-isothermal fluid flow. *Int. J. Numer. Methods Fluids* **18**, 127–162.

Bird, R. B., Armstrong, R. C. and Hassager, O., 1977. *Dynamics of Polymeric Fluids, Vol. 1: Fluid Mechanics*, 2nd edn, Wiley, New York.

Mitsoulis, E., Valchopoulos, J. and Mirza, F. A., 1985. A numerical study of the effect of normal stresses and elongational viscosity on entry vortex growth and extrudate swell. *Poly. Eng. Sci.* **25**, 677–669.

Pearson, J. R. A., 1979. Polymer flows dominated by high heat generation and low heat transfer. *Polym. Eng. Sci.* **18**, 1148–1154.

Petera, J. and Nassehi, V., 1993. Flow modelling using isoparametric Hermite elements. In: Taylor C. (ed.), *Numerical Methods in Laminar and Turbulent Flow*, Vol. VIII, Part 2, Pineridge Press, Swansea.

Petera, J. and Pittman, J. F. T 1994. Isoparametric Hermite elements', *Int. J. Num. Methods Eng.* **37**, 3489–3519.

Pittman, J. F. T. and Nakazawa, S., 1984. Finite element analysis of polymer processing operations. In: Pittman, J. F. T., Zienkiewicz, O. C., Wood, R. D. and Alexander, J. M. (eds), *Numerical Analysis of Forming Processes*, Wiley, Chichester.

Tanner, R. I., 1985. *Engineering Rheology*, Clarendon Press, Oxford.

Zienkiewicz, O. C. and Wu, J., 1991. Incompressibility without tears – how to avoid restrictions on mixed formulations. *Int. J. Numer. Methods Eng.* **32**, 1189–1203.

5

Rational Approximations and Illustrative Examples

In this chapter, selected examples of the application of weighted residual finite element schemes to the solution of polymer flow problems are presented. These examples provide a relevant background for description of the rational approximations that may be used to obtain realistic computer simulations for industrial polymer flow processes. In general, extension of the finite element techniques, described in the previous chapters, to a direct three-dimensional solution of non-Newtonian flow problems does not introduce any fundamental difficulties. But because of high cost of the required computations (or excessive length of execution times) a practical difficulty arises. Therefore rational approximations that reduce the 'size' of a problem without neglecting its core characteristics are of utmost importance in the development of practical computer models. The main categories of approximations used in the modelling of polymer processes are listed in Chapter 2. The techniques of implementation of these approximations are discussed in the following sections.

5.1 MODELS BASED ON SIMPLIFIED DOMAIN GEOMETRY

Straightforward one- or two-dimensional modelling of polymeric flow systems that correspond to a linear or planar geometry often yields useful results and is considered to be a permissible approximation in most cases. Important examples of such systems are found in fibre spinning and film casting which can be satisfactorily defined using one- or two-dimensional frameworks. Therefore the modelling of these processes is usually based on one- or two-dimensional schemes (e.g. see Andre *et al.*, 1998; Beaulne and Mitsoulis, 1999). Similarly if the process of interest, within a three-dimensional domain, is confined to a dominant direction (or cross-section) a reduction in model dimensionality may also be considered. Dispersive mixing of rubber/carbon compounds in partially filled internal mixers provides such an example in which the process of interest (i.e. breakdown of filler agglomerates) occurs predominantly in the cross-sectional plane of the rotor blades. Simulation of this process is considered next.

5.1.1   Modelling of the dispersion stage in partially filled batch internal mixers

A full account of finite element modelling of dispersive mixing can be found in the literature (see Nassehi and Ghoreishy, 1997; Nassehi *et al.*, 1997; Nassehi and Ghoreishy, 1998; Nassehi and Ghoreishy, 2001) and is not repeated in this section. The main focus here is to show that, despite the complexity of the process, two-dimensional models that incorporate its significant features can generate useful predictive results.

  Batch internal mixers are used extensively by industry to mix polymers with other material to produce composites with desirable properties. For example, mixing of rubber with carbon black is almost exclusively carried out in partially filled internal mixers. Essentially this process consists of three stages of incorporation, dispersion and distribution. During the incorporation stage pelletized carbon black and rubber are brought together, allowing the diffusion of macromolecular chains of the polymer into the void spaces inside the filler agglomerates. After this stage, dispersive mixing starts in which, through the imposition of an uneven stress field by the action of rotor blades, carbon black agglomerates are broken into smaller aggregates and dispersed within the matrix. Finally, the compound is distributed within the chamber to achieve uniformity. The described incorporation and dispersion phases mainly take place in the rotor blade cross-sectional plane and are stress dominated. In contrast, distribution of the material inside the mixer is pressure dominated and mainly takes place along the rotor axes (Clarke and Freakley, 1995). Therefore in the analysis of rubber mixing in internal mixers, it is reasonable to develop separate models for each stage of the process. In addition, the stress field that gives rise to the incorporation and dispersion of the phases, is obtainable from the simulation of the two-dimensional flow in the plane of the rotor blades cross-section. Therefore the development of two-dimensional models for this process is explained in the following section.

Flow simulation in a single blade partially filled mixer

The plane of the rotor blade cross-section representing the flow field configuration at the start of mixing in a partially filled single-blade mixer is shown in Figure 5.1. Initial distribution of the compound inside the mixer chamber corresponds to a fill factor of 71 per cent and is chosen arbitrarily. It is evident that the flow field within this domain should be modelled as a free surface regime with random moving boundaries. Available options for the modelling of such a flow regime are explained in Chapter 3, Section 5. In this example, utilization of the volume of fluid (VOF) approach based on an Eulerian framework is described. To maintain simplicity we neglect elastic effects and the variations of compound viscosity with mixing, and focus on the simulation of the flow corresponding to a generalized Newtonian fluid. In the VOF approach

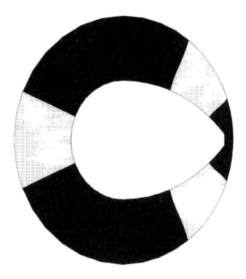

Figure 5.1   Initial configuration in a partially filled single blade internal mixer

the flow regime is treated as a multi-phase system in which compound- and air-filled regions are each assumed to represent a different phase. Similar to the compound-filled sections, the flow regime in the air-filled regions is considered to be incompressible. The exception is the narrow gap between the blade tip and the chamber wall, where an air bubble may become compressed if pressure exceeds beyond a given threshold. This is a realistic assumption because the chamber in an industrial mixer is not airtight and only air pockets trapped inside the compound matrix are expected to be compressed under high pressure.

Utilizing the domain symmetry of a single-blade mixer, it can be assumed that the blade remains stationary throughout the mixing and the flow is generated by the rotation of the chamber wall. This is readily achieved by the imposition of the appropriate boundary conditions and results in a significant simplification in the modelling by allowing the use of a constant finite element mesh for the entire simulation.

Figure 5.2 shows the finite element mesh corresponding to the configuration shown in Figure 5.1. This mesh consists of 225 nine-node bi-quadratic elements and its utilization in the present model is based on the application of iso-parametric mapping, described in Chapter 2.

The governing equations used in this case are identical to Equations (4.1) and (4.4) describing the creeping flow of an incompressible generalized Newtonian fluid. In the air-filled sections if the pressure exceeds a given threshold the equations should be switched to the following set describing a compressible flow

$$\frac{\partial \rho}{\partial t} + \frac{\partial}{\partial x}(\rho v_x) + \frac{\partial}{\partial y}(\rho v_y) = 0 \qquad \text{(continuity)} \qquad (5.1)$$

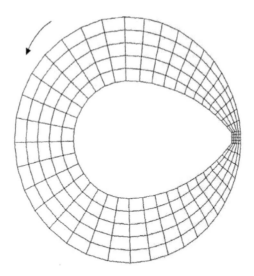

Figure 5.2   The mesh used in this example consisting of 225 nine-node bi-quadratic
elements

$$
\begin{cases}
\rho \dfrac{Dv_x}{Dt} = -\dfrac{\partial p}{\partial x} + \dfrac{\partial \tau_{xx}}{\partial x} + \dfrac{\partial \tau_{yx}}{\partial y} \\[4mm]
\rho \dfrac{Dv_y}{Dt} = -\dfrac{\partial p}{\partial y} + \dfrac{\partial \tau_{xy}}{\partial x} + \dfrac{\partial \tau_{yy}}{\partial y}
\end{cases}
\qquad \text{(motion)} \qquad (5.2)
$$

where

$$
\begin{cases}
\tau_{xx} = \mu \left[2\dfrac{\partial v_x}{\partial x} - \dfrac{2}{3}\left(\dfrac{\partial v_x}{\partial x} + \dfrac{\partial v_y}{\partial y} \right) \right] + k\left(\dfrac{\partial v_x}{\partial x} + \dfrac{\partial v_y}{\partial y} \right) \\[4mm]
\tau_{yy} = \mu \left[2\dfrac{\partial v_y}{\partial y} - \dfrac{2}{3}\left(\dfrac{\partial v_x}{\partial x} + \dfrac{\partial v_y}{\partial y} \right) \right] + k\left(\dfrac{\partial v_x}{\partial x} + \dfrac{\partial v_y}{\partial y} \right) \\[4mm]
\tau_{xy} = \tau_{yx} = \mu \left(\dfrac{\partial v_x}{\partial y} + \dfrac{\partial v_y}{\partial x} \right)
\end{cases}
\qquad (5.3)
$$

where μ and k are shear and volume (bulk) viscosity of the fluid, respectively.

The shear-dependent viscosity of the compound is found using the temperature-dependent form of the Carreau equation, described in Chapter 1, given as

$$
\eta = \frac{\eta_0}{\left[1 + (\lambda \dot{\gamma})^2\right]^{(1-n)/2}} \, e^{-b(T - T_{\text{ref}})}
\qquad (5.4)
$$

Temperature variations are found by the solution of the energy equation. The finite element scheme used in this example is based on the implicit θ time-stepping/continuous penalty scheme described in detail in Chapter 4, Section 5.

As described in Chapter 3, Section 5.1 the application of the VOF scheme in an Eulerian framework depends on the solution of the continuity equation for the free boundary (Equation (3.69)) with the model equations. The developed algorithm for the solution of the described model equations and updating of the free surface boundaries is as follows:

- Step 1 – the domain of interest is discretized into a mesh of finite elements.

- Step 2 – an initial configuration representing the partially filled discretized domain is considered and an array consisting of the appropriate values of F = 1, 0.5 and 0 for nodes containing fluid, free surface boundary and air, respectively, is prepared. The sets of initial values for the nodal velocity, pressure and temperature fields in the solution domain are assumed and stored as input arrays. An array containing the boundary conditions along the external boundaries of the solution domain is prepared and stored.

- Step 3 – the time variable is updated, incrementing it by Δt. To maintain the accuracy a small time step of 0.01 s is used.

- Step 4 – it is initially assumed that the flow field in the entire domain is incompressible and using the initial and boundary conditions the corresponding flow equations are solved to obtain the velocity and pressure distributions. Values of the material parameters at different regions of the domain are found via Equation (3.70) using the 'pseudo-density' method described in Chapter 3, Section 5.1.

- Step 5 – the obtained velocity field is used to solve the energy equation (see Chapter 3, Section 3).

- Step 6 – compound viscosity is updated via Equation (5.4).

- Step 7 – steps 4 to 6 are iterated until the solution is converged. Air-filled regions where the calculated pressure is more than 1 MPa are identified. Following the identification of high-pressure air-filled regions the solution is repeated by treating the air within these areas as compressible. Note that in a domain divided into compressible and incompressible parts the boundary line integrals, obtained by the application of Green's theorem to the second order derivatives in the equation of motion, will not be compatible at all inter-element boundaries. Therefore it cannot be assumed that all such terms will be automatically cancelled during the assembly of elemental stiffness equations. To avoid this difficulty an approximation based on the use of old time step values of the field variables in the line integral terms along the shared boundary of the compressible and incompressible parts is introduced. To maintain accuracy of the solutions in the VOF method small time steps should be used, hence the effect of this approximation is insignificant.

- Step 8 – the new values of the nodal velocities found at the end of step 7 are used as input and the free surface equation is solved.

- Step 9 – using updated values of the free surface function the location of the free surfaces are identified and the positions of each phase in the current flow domain are marked accordingly.

- Step 10 – steps 3 to 9 are repeated and the solution is advanced in time until the required end of the simulation.

The predicted free boundary distributions within a chamber of 0.05 m radius for a fluid with the following physical parameters, $\eta_0 = 10^5$, $n = 0.25$, $b = 0.014$, $T_{ref} = 373$, $\lambda = 1.5$, $\rho = 1055$, $c = 1255$, $k = 0.13$ (SI units), used in the Carreau, flow and energy equations, and penalty parameter of 10^9 after 10, 30, 50 and 100 time steps are shown in Figures 5.3a to 5.3d.

Flow simulation in a partially filled twin blade mixer

Simplification achieved by using a constant mesh in the modelling of the flow field in a single-blade mixer is not applicable to twin-blade mixers. Although the model equations in both simulations are identical the solution algorithm for twin-blade mixers cannot be based on the VOF method on a fixed domain and instead the Arbitrary Lagrangian–Eulerian (ALE) approach, described in Chapter 3, Section 5.2, should be used. However, the overall geometry of the plane of the rotors blades cross-section is known and all of the required mesh configurations can be generated in advance and stored in a file to speed up the calculations. Figure 5.4 shows the finite element mesh corresponding to 19 successive time steps from the start of the simulation in a typical twin-blade tangential rotor mixer. The finite element mesh configurations correspond to counter-rotating blades with unequal rotational velocities set to generate an uneven stress field for enhancing dispersive mixing efficiency. Calculation of mesh velocity, required for modification of the free surface equation (see Equation (3.73)) at each time step, is based on the following equations (Ghoreishy, 1997)

$$\begin{cases} v_{mx} = \dfrac{(x)_n^t - (x)_n^{t-\Delta t}}{\Delta t} \\[2mm] v_{my} = \dfrac{(y)_n^t - (y)_n^{t-\Delta t}}{\Delta t} \end{cases} \qquad (5.5)$$

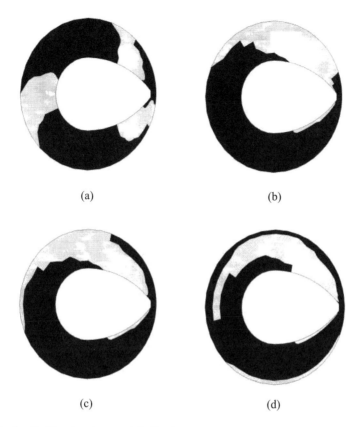

(a)   (b)

(c)   (d)

Figure 5.3   (a–d) Simulated material distribution within the single-blade mixer after 10, 30, 50 and 100 time steps

where $(x)_n^t$ etc. are the coordinates of node n at time t. As already mentioned the time increment used in the VOF method should be small and hence Equation (5.5), which represents linear nodal movements between successive mesh configurations, does not generate inaccurate results. In this example, the solution algorithm starts by inserting zero for mesh velocity in the governing equations. These equations are then solved in the first mesh corresponding to the initial configuration inside the mixer. After the convergence of the iterations this solution yields nodal velocities (and other field variables) at the end of the first time step. After this stage, mesh velocities are found using Equation (5.5) and the solution proceeds to the next step. The procedure is repeated until a complete set of field data for all of the configurations corresponding to different relative positions of rotor blades is found.

Initial distribution and the predicted free surface boundaries within the twin-blade mixer represented by the mesh configurations shown in Figure 5.4, after 30, 60 and 90° rotation of the left blade are presented in Figures 5.5a to 5.5d, respectively. Samples of the predicted velocity fields after 30 and 45° rotation of the left rotor are shown in Figures 5.6a to 5.6b, respectively. The finite element

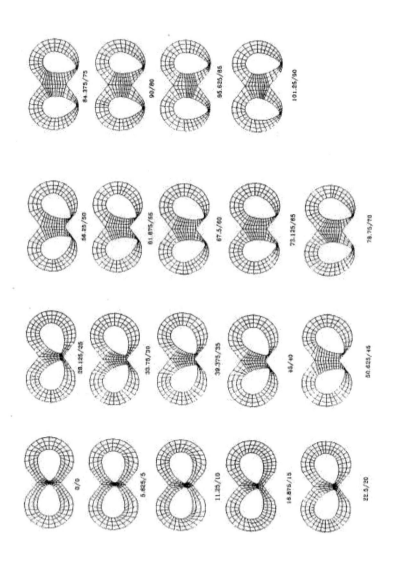

Figure 5.4 The finite element mesh configurations in the Arbitrary Lagrangian–Eulerian scheme

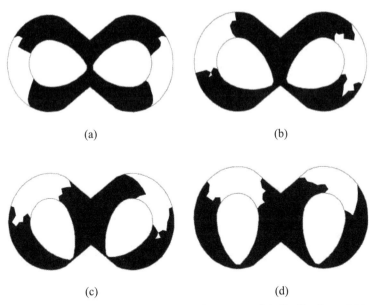

(a)   (b)

(c)   (d)

Figure 5.5 (a–d) Initial configuration and simulated material distribution within the twin-blade mixer after 30, 60 and 90° rotation of the left blade

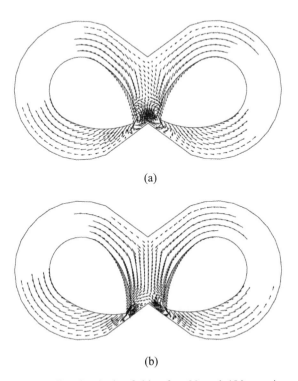

(a)

(b)

Figure 5.6 (a,b) The predicted velocity fields after 30 and 45° rotation of the left blade in the partially filled twin-blade mixer

scheme used in this simulation is based on the implicit θ/continuous penalty method and the physical parameters are equal to the values given in the previous example.

5.2 MODELS BASED ON SIMPLIFIED GOVERNING EQUATIONS

As discussed in the previous chapters, utilization of viscoelastic constitutive equations in the finite element schemes requires a significantly higher computational effort than the generalized Newtonian approach. Therefore an important simplification in the model development is achieved if the elastic effects in a flow system can be ignored. However, almost all types of polymeric fluids exhibit some degree of viscoelastic behaviour during their flow and deformation. Hence the neglect of these effects, without a sound evaluation of the flow regime characteristics, which may not allow such a simplification, can yield inaccurate results.

In the following sections modelling of the free surface flow of silicon rubber in a two-dimensional domain using both generalized Newtonian and viscoelastic constitutive equations is presented. Comparing the results of these simulations, the effects of assuming a simplified generalized Newtonian behaviour against using a viscoelastic constitutive equation for this fluid are evaluated. This evaluation illustrates the conditions under which the described simplification leads to useful predictions and can hence be regarded as acceptable.

The design and operation of a flow visualization system for highly viscous fluids, such as silicon rubber, has been reported by Ghafouri and Freakley (1994). This system consists mainly of a rotating roll and fixed-blade assembly, as is shown in Figure 5.7, and can be used to generate and maintain, essentially,

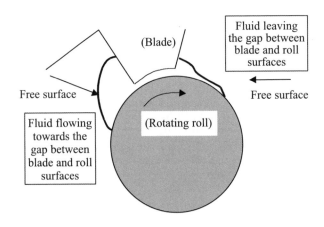

Figure 5.7 Schematic diagram of the modelled flow visualization rig

two-dimensional Couette flow regimes. To generate a Couette flow, a given mass of fluid is placed (or banded) in front of the fixed blade. As the roll starts to rotate the fluid is dragged through the gap between the blade and roll surfaces and a transient reservoir in front of the blade is formed. The main flow domain is hence defined as a converging channel connecting two free surface flow regions on its sides through a narrow gap. Published flow visualization results showing the evolution of the free surface boundaries at the sides of this channel provide experimental data for evaluation of the accuracy of the simulations obtained using various constitutive equations.

5.2.1   Simulation of the Couette flow of silicon rubber – generalized Newtonian model

Assuming a generalized Newtonian behaviour for silicon rubber, the Couette flow established in the described flow visualization experiment can be modelled using the VOF scheme based on a fixed (Eulerian) framework. The governing equations and finite element discretization in this model are identical to the implicit θ/continuous penalty scheme explained in Chapter 4, Section 5. This scheme is used in conjunction with the VOF procedure described in Chapter 3, Section 5.1. Figure 5.8 shows the finite element mesh corresponding to the maximum extent that the fluid can be expected to reach during the simulation time of 60 s for a roll velocity of 1 rpm. Predicted free surface boundaries at two different times are compared with the experimental data (Nassehi and Ghoreishy, 1997) in Figures 5.9a to 5.9b. The close comparison between the simulation results and the experimental data points to the conclusion that despite ignoring the elastic effects the simulated flow domain is realistic. The diminished extent of swelling at the exit from the narrow gap is attributed to the reduction in the influence of normal stresses in the present shear-induced flow regime.

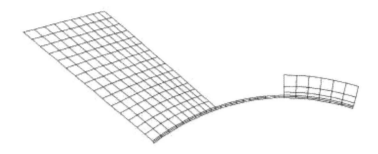

Figure 5.8   The finite element mesh used to model free surface flow in example

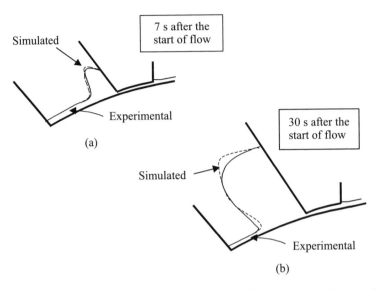

Simulated

7 s after the
start of flow

Experimental

(a)

30 s after the
start of flow

Simulated

Experimental

(b)

Figure 5.9   (a,b) Comparison of the simulated and observed free surface positions

5.2.2   Simulation of the Couette flow of silicon rubber – viscoelastic model

Keeping all of the flow regime conditions identical to the previous example, we now consider a finite element model based on treating silicon rubber as a viscoelastic fluid whose constitutive behaviour is defined by the following upper-convected Maxwell equation

$$
\begin{cases}
\lambda\left(\dfrac{\partial T_{xx}}{\partial t} + v_x\dfrac{\partial T_{xx}}{\partial x} + v_y\dfrac{\partial T_{xx}}{\partial y} - 2\dfrac{\partial v_x}{\partial x}T_{xx} - 2\dfrac{\partial v_x}{\partial y}T_{xx}\right) + T_{xx} = \dfrac{\eta}{\lambda} \\[3mm]
\lambda\left(\dfrac{\partial T_{yy}}{\partial t} + v_x\dfrac{\partial T_{yy}}{\partial x} + v_y\dfrac{\partial T_{yy}}{\partial y} - 2\dfrac{\partial v_y}{\partial x}T_{xy} - 2\dfrac{\partial v_x}{\partial y}T_{yy}\right) + T_{yy} = \dfrac{\eta}{\lambda} \\[3mm]
\lambda\left(\dfrac{\partial T_{xy}}{\partial t} + v_x\dfrac{\partial T_{xy}}{\partial x} + v_y\dfrac{\partial T_{xy}}{\partial y} - \dfrac{\partial v_y}{\partial x}T_{xx} - \dfrac{\partial v_x}{\partial y}T_{yy}\right) + T_{xy} = 0
\end{cases} \quad (5.6)
$$

where T_{xx} etc. are the components of Maxwell stress defined as

$$
T = \frac{\eta}{\lambda}\delta + \tau \quad (5.7)
$$

where η is viscosity, λ is relaxation time and δ is the Kronecker delta. The upper-convected Maxwell constitutive model is written as Equation (5.6) for the advantage that the simplified representation of the right-hand side provides in a numerical solution. Substitution of Equation (5.7) into Equation (5.6) results in the more familiar form of the Maxwell model given in Chapter 1.

Equation (5.6) is a hyperbolic partial differential equation and its finite element solution in a fixed coordinate system requires the use of an upwinding scheme (see Chapter 3, Section 2). However, numerical damping generated by upwinding may adversely affect the accuracy of the solution. Therefore it is preferable to use a Lagrangian framework for the solution of free surface and constitutive equations in this problem. Equations of continuity and motion, on the other hand, can be solved using a fixed coordinate system. An optimum solution algorithm based on a decoupled procedure, in which flow equations are solved after the updating of the free surface and the calculation of stress components, is thus developed. The Taylor–Galerkin/U–V–P scheme corresponding to a slightly compressible continuity equation for viscoelastic fluids, described in Chapter 4, Section 5, is used to solve the flow part of the model. The constitutive and free surface equations are solved at a different step, using the VOF method in a Lagrangian framework. Details of the algorithm for the updating of free boundaries is given in Chapter 3, Section 5.3, here we focus on the solution of the constitutive equation in a moving coordinate system. Similar to the solution of the free surface equation, the moving system in this problem is constructed along the fluid particle trajectories. This framework can be defined as a general curvilinear system in which a line element is given by the 'metric form' (Spiegel, 1974), as

$$[d\underline{\tilde{X}}(t)]^2 = \gamma^{pq}(t)\ dl_p dl_q \tag{5.8}$$

where summation over repeated indices is assumed and

$$\gamma^{pq} = \frac{\partial x^p}{\partial X^k}\frac{\partial x^q}{\partial X^m}g^{km} \tag{5.9}$$

where g^{km} are the components of the metric of the reference coordinate system (in a Cartesian coordinate system $g^{km} = \delta^{km}$). Components of the stress tensor are written as

$$\Pi^{pq} = \frac{\partial l^p}{\partial X^m}T^{km}\frac{\partial l^q}{\partial X^k} = \frac{\partial x^p}{\partial X^m}T^{mk}\frac{\partial x^q}{\partial X^k} \tag{5.10}$$

In the moving coordinate system the constitutive equation (i.e. Equation (5.6)) can be written as (Bird *et al.*, 1977)

$$\lambda\dot{\Pi}^{pq} + \Pi^{pq} = \frac{\eta}{\lambda}\gamma^{pq} \tag{5.11}$$

Note that in a Lagrangian system the convection terms in Equation (5.11) vanish.

Application of the previously described θ time-stepping method to Equation (5.11) gives

$$(\lambda + \theta \Delta t)\Pi^{pq}\Big|^{n+1} = [\lambda - (1-\theta)\Delta t]\Pi^{pq}\Big|^n + \theta \frac{\eta}{\lambda}\gamma^{pq}\Big|^{n+1} + (1-\theta)\frac{\eta}{\lambda}\gamma^{pq}\Big|^n \qquad (5.12)$$

After rearranging, the Galerkin-weighted residual statement arising from Equation (5.12) can be written as

$$\int\limits_{\Omega_{n+1}} \lambda N_I \overline{\Pi}^{pq}\mathrm{d}\Omega_{n+1} - \int\limits_{\Omega_n} \lambda N_I \overline{\Pi}^{pq}\mathrm{d}\Omega_n = \int\limits_{t_n}^{t_{n+1}} \int\limits_{\Omega_t} N_I\left(-\overline{\Pi}^{pq} + \frac{\eta}{\lambda}\gamma^{pq}\right)\mathrm{d}\Omega_t \mathrm{d}t \qquad (5.13)$$

where an over bar means elemental discretization in the usual manner. The working equation yielding the stress components in the current time level of $(n + 1)$ can now be constructed, combining Equations (5.12) and (5.13), as

$$(M_{IJ}^{n+1} + \theta \Delta t K_{IJ}^{n+1})(\overline{\Pi}_J^{pq})^{n+1} = [M_{IJ}^n - (1-\theta)\Delta t K_{IJ}^n](\overline{\Pi}_J^{pq})^n + \Delta t B_I^{n+\theta} \qquad (5.14)$$

where

$$\begin{cases} M_{IJ}^{n+1} = \int\limits_{\Omega_{n+1}} \lambda N_I N_J \mathrm{d}\Omega_{n+1} \\[2mm] M_{IJ}^n = \int\limits_{\Omega_n} \lambda N_I N_J \mathrm{d}\Omega_n \\[2mm] K_{IJ}^{n+1} = \int\limits_{\Omega_{n+1}} N_I N_J \mathrm{d}\Omega_{n+1} \\[2mm] K_{IJ}^n = \int\limits_{\Omega_n} N_I N_J \mathrm{d}\Omega_n \end{cases} \qquad (5.15)$$

and

$$B_I^{n+\theta} = \theta \int\limits_{\Omega_{n+1}} N_I \frac{\eta}{\lambda}\gamma^{pq}\mathrm{d}\Omega_{n+1} + (1-\theta)\int\limits_{\Omega_n} N_I \frac{\eta}{\lambda}\gamma^{pq}\mathrm{d}\Omega_n \qquad (5.16)$$

Note that the shape functions used in the above discretization preserve their originally defined forms. This is in contrast to the Lagrangian formulations in which the shape functions need to be modified (Donea and Quartapelle, 1992).

In Equation (5.14), $(\overline{\Pi}_J^{pq})^n$ is found by interpolating existing nodal values at the old time step and then transforming the found value to the convected coordinate system. Calculation of the components of γ^{pq} and $(\overline{\Pi}_J^{pq})^n$ depends on the evaluation of first-order derivatives of the transformed coordinates (e.g. as seen in Equation (5.9)). This gives the measure of deformation experienced by the fluid between time steps of n and $n + 1$. Using the time-independent local coordinates of a fluid particle (ξ, η) we have

$$\frac{\partial x^p}{\partial X^k} = \frac{\partial x^p}{\partial \xi}\frac{\partial \xi}{\partial X^k} + \frac{\partial x^p}{\partial \eta}\frac{\partial \eta}{\partial X^k} \tag{5.17}$$

Smoothness of the transformation between the fixed and moving systems depends on the inter-element continuity of first-order derivatives of coordinates. This can be achieved by using C^1 continuous Hermite elements. As for all types of finite elements, applicability of undistorted Hermite elements in practical problems is very limited. Therefore a procedure for the construction of isoparametric C^1 continuous Hermite elements is incorporated in the present solution scheme. Details of a numerical technique for the construction of isoparametric C^1 continuous Hermite elements can be found in the literature (e.g. see Petera and Pittman, 1994). This method is used to obtain a discretization for the working equation (5.14) based on the isoparametric form of the rectangular Bogner–Fox–Schmit element, shown in Figure 2.10. Results obtained using the described viscoelastic model are only marginally different from the results shown in Figures 5.9a and 5.9b (Petera and Nassehi, 1996). Although this confirms the overall validity of the utilization of a simplified constitutive equation for this flow process, detailed and direct comparison of the two sets of results is needed to obtain a quantitative evaluation of the effects of the adopted approximation. To carry out such an evaluation let us consider the simulated velocity fields at the exit section of a domain analagous to the narrow gap in the converging flow channel between the roll and blade shown in Figure 5.7. Figures 5.10a and 5.10b show the velocity field corresponding to the generalized Newtonian and

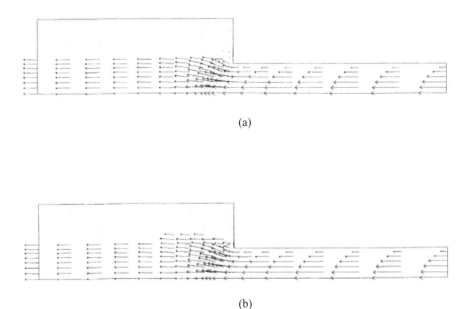

(a)

(b)

Figure 5.10   (a,b) Comparison of the simulated die swell in a Couette flow for the power-law and upper-convected Maxwell fluids

viscoelastic models, respectively. All other data used to obtain these results are identical. As this comparison shows the overall velocity fields in the two cases are not significantly different. However, the distinct 'localized' swelling, obtained using the viscoelastic constitutive equation, has not been generated by the generalized Newtonian model. It is therefore evident that the latter model has a restricted applicability and should only be used in situations where localized effects resulting from the viscoelasticity of the fluid can be ignored.

5.3   MODELS REPRESENTING SELECTED SEGMENTS OF A LARGE DOMAIN

Under certain conditions it may be appropriate to focus on the modelling of a segment of a larger domain in order to obtain detailed results within that section, while maintaining computing economy. To develop such a model, first the entire flow domain is simulated using a relatively coarse finite element mesh. The results generated at the end of this stage are used to define the boundary conditions at the borders of the selected part. This section is then modelled utilizing a refined mesh to obtain detailed predictions about phenomena of interest in the flow field. The procedure can be repeated using a step-by-step approach in which the zooming on a segment within a very large domain is achieved through successive reduction of the size of the segments at each step of modelling.

In the following sections examples of the application of this procedure to the analysis of specific phenomena such as wall slip and stress overshoot which affect polymeric flow processes are illustrated.

5.3.1   Prediction of stress overshoot in the contracting sections of a symmetric flow domain

It has been long established that during the flow of long-chain polymers stresses within the fluid may rise significantly (overshoot) at certain locations in the domain. This phenomenon is more clearly observed in contracting flows where the polymer moves from wider to narrower sections of the domain. Let us consider the flow of a polymeric fluid in a two-dimensional gap between the cross-sectional plane of a symmetric screw and a cylindrical outer tube. Initially the entire domain is discretized into a mesh consisting of 512 bi-quadratic finite elements, as shown in Figure 5.11. The flow inside this domain is generated by rotation of the screw in the anticlockwise direction. Using the simulation results obtained on this mesh, the boundary conditions for a representative section of the domain confined between two flights are defined and evaluated. A more refined mesh consisting of 256 bi-quadratic elements for this section is constructed. Considering the symmetry of the domain, simulation of stress field in this section should provide an insight for the entire domain. Predicted normal

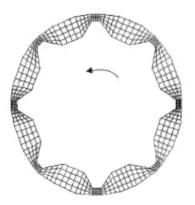

Figure 5.11   Finite element discretization of the symmetric domain in example 5.3.1

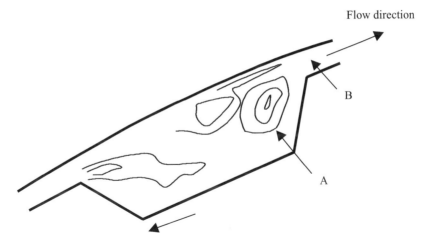

Figure 5.12   Schematic diagram of the predicted normal stress contours in a typical section of the symmetric domain shown in Figure 5.11

stress contours corresponding to a peak at the entrance to the narrow gap between the flight and the outside wall are shown, schematically, in Figure 5.12.

Solution of the flow equations has been based on the application of the implicit θ time-stepping/continuous penalty scheme (Chapter 4, Section 5) at a separate step from the constitutive equation. The constitutive model used in this example has been the Phan-Thien/Tanner equation for viscoelastic fluids given as Equation (1.27) in Chapter 1. Details of the finite element solution of this equation are published elsewhere and not repeated here (Hou and Nassehi, 2001). The predicted normal stress profiles along the line AB (see Figure 5.12) at five successive time steps are shown in Figure 5.13. The predicted pattern is expected to be repeated throughout the entire domain.

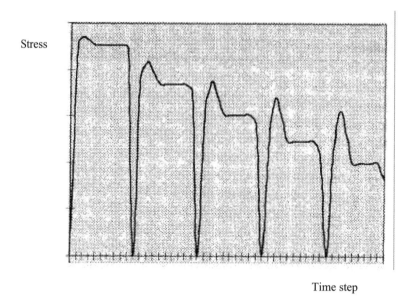

Stress

Time step

Figure 5.13   The predicted pattern of the stress overshoot in example

5.3.2   Simulation of wall slip in a rubber mixer

Fluid slippage at solid wall surfaces is a common phenomenon affecting polymer flow regimes. In particular in processes involving long-chain elastomers,
such as natural or synthetic rubbers, a noticeable percentage of the energy input
may be wasted through wall slip in the domain. As explained in Chapter 3,
Section 4.2 the imposition of wall slip boundary conditions in a finite element
model should be based on Navier's slip condition. A straightforward method for
the imposition of these boundary conditions is to modify the original working
equations in a flow model via the incorporation of the discretized relationships
arising from the components of the Navier's slip condition (e.g. Equations (3.62)
and (3.63)). This is shown as:

Modified weighted residual statement = Original weighted residual statement +
G [slip-wall boundary conditions].

G is a multiplier which is zero at locations where slip condition does not apply
and is a sufficiently large number at the nodes where slip may occur. It is
important to note that, when the shear stress at a wall exceeds the threshold of
slip and the fluid slides over the solid surface, this may reduce the shearing to
below the critical value resulting in a renewed stick. Therefore imposition of wall
slip introduces a form of 'non-linearity' into the flow model which should be
handled via an iterative loop. The slip coefficient (i.e. β in the Navier's slip
condition given as Equation (3.59) is defined as

$$\beta = \frac{1}{k_s} \qquad\qquad (5.18)$$

where k_s is the experimentally determined coefficient of sliding friction between the fluid and a solid surface. The iteration loop in the imposition of slip condition may start by assuming $\beta = \beta_0 = 1/k_s$ and subsequently updating this coefficient as

$$\beta = \beta_0 \frac{|V|}{I_2(d)} \qquad\qquad (5.19)$$

where V is a characteristic flow velocity and $I_2(d)$ is the second invariant of the rate of deformation tensor. Equation (5.19) should be regarded as a heuristically defined relationship between the extent of wall slip at a given stage of flow and the state of stress at that time. In practice, comparison with experimental data may be needed to verify the accuracy of model predictions.

Predicted velocity fields in a segment adjacent to the tip of the blade in the single-blade mixer, described in a previous sub-section, before and after imposition of the wall slip are shown in Figures 5.14a and 5.14b, respectively. As expected, momentum transfer from the rotating wall to the fluid is significantly affected by the imposition of the wall slip. In Figures 5.15a and 5.15b temperature contours corresponding to these velocity fields are shown.

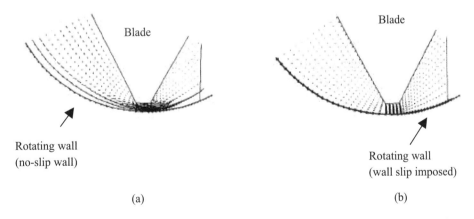

Figure 5.14   (a) The predicted velocity field corresponding to no-slip wall boundary conditions. (b) The predicted velocity field corresponding to partial slip boundary conditions

As comparison of the simulated temperature fields shows, fluid slippage results in temperature peak shifting from a location furthest away from the

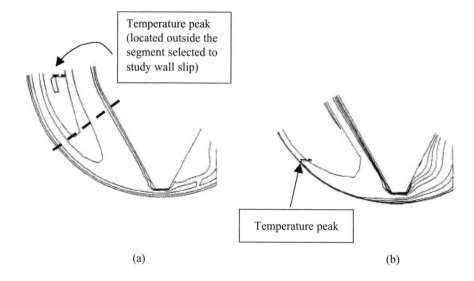

Temperature peak
(located outside the
segment selected to
study wall slip)

Temperature peak

(a)   (b)

Figure 5.15   (a) The predicted temperature distribution corresponding to the no-slip
conditions. (b) The predicted temperature distribution corresponding to the
partial slip conditions

cooling walls to a layer next to the rotating surface. This provides an indication
that because of the wall slip the energy input is converted to heat instead of
generating flow.

5.4   MODELS BASED ON DECOUPLED FLOW EQUATIONS – SIMULATION OF THE FLOW INSIDE A CONE-AND-PLATE RHEOMETER

In Chapter 4 the development of axisymmetric models in which the radial and
axial components of flow field variables remain constant in the circumferential
direction is discussed. In situations where deviation from such a perfect symmetry is small it may still be possible to decouple components of the equation
of motion and analyse the flow regime as a combination of one- and two-
dimensional systems. To provide an illustrative example for this type of
approximation, in this section we consider the modelling of the flow field inside
a cone-and-plate viscometer.

Rotating cone viscometers are among the most commonly used rheometry
devices. These instruments essentially consist of a steel cone which rotates in a
chamber filled with the fluid generating a Couette flow regime. Based on the
same fundamental concept various types of single and double cone devices are
developed. The schematic diagram of a double cone viscometer is shown in

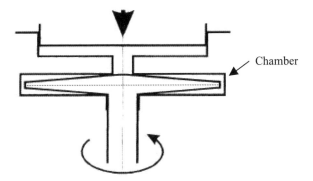

Figure 5.16 Schematic diagram of a bi-conical cone-and-plate viscometer

Figure 5.16. It is assumed that by using an exactly symmetric cone a shear rate distribution, which is very nearly uniform, within the equilibrium (i.e. steady state) flow field can be generated (Tanner, 1985). Therefore in this type of viscometry the applied torque required for the steady rotation of the cone is related to the uniform shearing stress on its surface by a simplified theoretical equation given as

$$\psi_{\text{exp}} = \psi_{\text{the}} = \tfrac{2}{3}\pi\dot{\gamma}\eta(\dot{\gamma})R_0^3 \tag{5.20}$$

where ψ_{exp} and ψ_{the} are the experimental and theoretical torque, respectively. The shear rate is given by

$$\dot{\gamma} = \frac{\omega}{\alpha_0} \tag{5.21}$$

where ω is the angular velocity of rotation. The geometrical parameters given as R_0 and α_0 are the cone radius and cone angle (neglecting the shaft thickness), respectively. The material parameters calculated on the basis of viscosity measured by cone-and-plate rheometry are only true if the stress on the cone surface and the temperature field in the flow domain remain exactly uniform during the experiment. In-depth analysis, however, reveals that distinctly non-uniform stress fields can develop inside these devices (Chaturani and Narasimman, 1990). One reason for this deviation from uniformity is that the flow near the outer wall of the chamber is not exactly unidirectional and uniform. But it has been shown that the maximum error from this source is less than five percent (Kaye et al., 1968). More important reasons for the deviation of the cone-and-plate experiment from a theoretical 'viscometric' flow are due to the action of centrifugal forces, which cause secondary flows, and the non-uniformity of viscous heat generation within the flow domain. In experiments involving highly viscous fluids the non-uniformity of the rate of internal heat generation gives rise to a non-uniform stress distribution on the cone surface. This effect is more noticeable for Newtonian fluids. In relatively low viscosity fluids the secondary flows are the main cause of the stress non-uniformity in the cone-and-plate flow domain. This type of non-uniformity is more significant in experiments involving non-Newtonian fluids. Unlike the wall effects, the experimental errors arising from these types of flow field non-uniformity cannot be readily estimated.

In the absence of direct experimental measurements, finite element modelling provides a reliable method for the evaluation of the errors resulting from the described stress non-uniformity in the cone-and-plate flow domain. Despite the deviation from an exactly viscometric regime, it is still possible to assume that the flow inside the cone-and-plate rheometer is very nearly axisymmetric. Therefore for generalized Newtonian fluids, the circumferential component of the equation of motion can be decoupled from the other two components without neglecting centrifugal and Coriolis forces acting upon the flow field. In viscoelastic fluids, however, even after assuming an axisymmetric flow regime inside the domain, some of the terms in the circumferential component of the equation of motion remain dependent on the variables in the radial and axial directions. In the following sections the governing equations and finite element modelling of the flow field inside a cone-and-plate viscometer are described.

5.4.1   Governing equations

For equilibrium (i.e. steady) flow in a cone-and-plate domain with perfect symmetry it is evident that any changes in the circumferential (i.e. θ) direction can be neglected. In this case the governing conservation equations describing the flow can be considerably simplified by omitting all terms containing derivatives with respect to the independent variable, θ. Note that the approach adopted for decoupling of the components of the governing equations in the present nearly axisymmetric flow is different from the procedure used for perfectly axisymmetric conditions. As described in Chapter 4, Section 1.3, in the latter case the governing equations can be analytically integrated with respect to θ. After omitting the derivatives with respect to θ the resulting system of the model equations in the (r, θ, z) coordinate system is written as:

- Continuity

$$\frac{\partial v_r}{\partial r} + \frac{v_r}{r} + \frac{\partial v_z}{\partial z} = 0 \tag{5.22}$$

- Component of the equation of motion in the predominant, i.e. θ direction

$$\rho\left(v_r\frac{\partial v_\theta}{\partial r} + v_z\frac{\partial v_\theta}{\partial z} + \frac{v_r v_\theta}{r}\right) = \frac{1}{r^2}\frac{\partial}{\partial r}(r^2\tau_{\theta r}) + \frac{\partial}{\partial z}(\tau_{\theta z}) \tag{5.23}$$

where v_r, v_z and v_θ are the components of the velocity vector and $\tau_{\theta r}$ and $\tau_{\theta z}$ are the extra stress components.

- Components of the equation of motion in the (r, z) plane are given as

$$\begin{cases} \rho\left(v_r\dfrac{\partial v_r}{\partial r} + v_z\dfrac{\partial v_r}{\partial z} - \dfrac{v_\theta^2}{r}\right) = -\dfrac{\partial p}{\partial r} + \dfrac{1}{r}\dfrac{\partial}{\partial r}(r\tau_{rr}) + \dfrac{\partial}{\partial z}(\tau_{rz}) - \dfrac{\tau_{\theta\theta}}{r} \\[4mm] \rho\left(v_r\dfrac{\partial v_z}{\partial r} + v_z\dfrac{\partial v_z}{\partial z}\right) = -\dfrac{\partial p}{\partial z} + \dfrac{1}{r}\dfrac{\partial}{\partial r}(r\tau_{rz}) + \dfrac{\partial}{\partial z}(\tau_{zz}) - \rho g \end{cases} \tag{5.24}$$

where p is pressure, g is acceleration due to gravity and τ_{rr}, τ_{rz}, $\tau_{\theta\theta}$ and τ_{zz} are the components of the extra stress tensor.

- Energy equation

$$\rho c\left(v_r\frac{\partial T}{\partial r} + v_z\frac{\partial T}{\partial z}\right) = \frac{1}{r}\frac{\partial}{\partial r}\left(kr\frac{\partial T}{\partial z}\right) + \frac{\partial}{\partial z}\left(k\frac{\partial T}{\partial z}\right) + \eta\dot{\gamma}^2 \tag{5.25}$$

where T is temperature, c is the specific heat capacity, k is the thermal conductivity and $(\eta\dot{\gamma}^2)$ is the viscous heat generation. The shear rate, $\dot{\gamma}$, is expressed as

$$\dot{\gamma} = \left[2\left(\frac{\partial v_r}{\partial r}\right)^2 + 2\left(\frac{\partial v_z}{\partial z}\right)^2 + \left(\frac{\partial v_r}{\partial z} + \frac{\partial v_z}{\partial r}\right)^2 + \left(\frac{\partial v_\theta}{\partial r} - \frac{v_\theta}{r}\right)^2 + \left(\frac{\partial v_\theta}{\partial z}\right)^2\right]^{\frac{1}{2}} \tag{5.26}$$

In order to account for the heat loss through the metallic body of the cone, a heat conduction equation, obtained by the elimination of the convection and source terms in Equation (5.25), should also be incorporated in the governing equations.

In the Couette flow inside a cone-and-plate viscometer the circumferential velocity at any given radial position is approximately a linear function of the vertical coordinate. Therefore the shear rate corresponding to this component is almost constant. The heat generation term in Equation (5.25) is hence nearly constant. Furthermore, in uniform Couette regime the convection term is also zero and all of the heat transfer is due to conduction. For very large conductivity coefficients the heat conduction will be very fast and the temperature profile will be uniform. Most polymers, however, have very small conductivities and the temperature profile will be parabolic with a maximum between moving and stationary surfaces (Bird *et al.*, 1959). This effect will be more significant in the bi-conical devices than a single cone rheometer because most of the heat transfer is expected to be through the body of the cone. This analysis shows that heat transport within the flow domain and the cone body can be modelled simultaneously:

- Constitutive equation.
  For a generalized Newtonian fluid the components of the extra stress and rate of deformation tensors in this domain are related by

$$
\begin{cases}
\tau_{\theta r}^{v} = \eta r \dfrac{\partial}{\partial r}\left(\dfrac{v_{\theta}}{r}\right) \\[4mm]
\tau_{\theta z}^{v} = \eta \dfrac{\partial}{\partial z}(v_{\theta})
\end{cases}
\tag{5.27}
$$

and

$$
\begin{cases}
\tau_{rr}^{v} = 2\eta \dfrac{\partial v_{r}}{\partial r} \\[4mm]
\tau_{zz}^{v} = 2\eta \dfrac{\partial v_{z}}{\partial z} \\[4mm]
\tau_{\theta\theta}^{v} = 2\eta \dfrac{v_{r}}{r} \\[4mm]
\tau_{rz}^{v} = \eta \left(\dfrac{\partial v_{r}}{\partial z} + \dfrac{\partial v_{z}}{\partial r}\right)
\end{cases}
\tag{5.28}
$$

The derivation of an appropriate constitutive equation for viscoelastic fluids in this flow field can be based on the upper-convected Maxwell equation (Equation (1.24). However, a crucial point to note is that the symmetry of the flow regime generated by the rotation of a very small angle cone is only slightly perturbed. Therefore the coupling (i.e. dependency) of the radial and axial field variables on the circumferential variables is weak. Based on this observation the components of the Maxwell equation can be expanded using a perturbation expansion (Nayfeh, 1993) and only keeping the significant terms in the model. Following this approach, the appropriate constitutive equation in the cone-and-plate domain is found by perturbation expansion of the components of the Maxwell equation, originally written in a cylindrical coordinate system, with respect to the small cone angle β (Olagunju and Cook, 1993). For a steady-state flow, keeping up to the second order-terms in the expansions, these components are written in terms of $\alpha = \tan(\beta)$, as

$$
\begin{cases}
\Sigma = 2\eta \dfrac{\partial u}{\partial r} - 2De\xi \dfrac{\partial u}{\partial h} + O(\alpha^2) \\[2ex]
\xi = \eta \dfrac{\partial u}{\partial h} + O(\alpha^3) \\[2ex]
\gamma = \eta \left(\dfrac{\partial w}{\partial r} - \dfrac{w}{r} \right) + De \left(\dfrac{\partial u}{\partial h} \Pi + \dfrac{\partial w}{\partial h} \xi \right) + O(\alpha^2) \\[2ex]
\Gamma = 2\eta \dfrac{\partial v}{\partial h} + O(\alpha^2) \\[2ex]
\Pi = \eta \dfrac{\partial w}{\partial h} + \alpha^2 De \left[-u \dfrac{\partial \Pi}{\partial r} - v \dfrac{\partial \Pi}{\partial h} + \dfrac{\partial v}{\partial h} \Pi + \left(\dfrac{\partial w}{\partial r} - \dfrac{w}{r} \right) \xi + \dfrac{\partial w}{\partial h} \Gamma + \dfrac{u}{r} \Pi \right] \\[2ex]
\Delta = 2De \dfrac{\partial w}{\partial h} \Pi + \alpha^2 De \left[-u \dfrac{\partial \Delta}{\partial r} - v \dfrac{\partial \Delta}{\partial h} + 2\gamma \left(\dfrac{\partial w}{\partial r} - \dfrac{w}{r} \right) + 2 \dfrac{u}{r} \Delta \right]
\end{cases}
\tag{5.29}
$$

where

$$
\begin{cases}
\Sigma = \dfrac{\tau_{rr}}{\alpha^2}, \quad \xi = \dfrac{\tau_{rz}}{\alpha}, \quad \gamma = \tau_{r\theta} \\[2ex]
\Gamma = \dfrac{\tau_{zz}}{\alpha^2}, \quad \Pi = \alpha \tau_{z\theta}, \quad \Delta = \alpha^2 \tau_{\theta\theta} \\[2ex]
u = \dfrac{v_r}{\alpha^2}, \quad v = \dfrac{v_z}{\alpha^3}, \quad w = v_\theta, \quad h = \dfrac{z}{\alpha}
\end{cases}
$$

and $De = t_{ref}(\omega)$, where t_{rel} is the Maxwell relaxation time and ω is the rotational speed.

A similar approximation should be applied to the components of the equation of motion and the significant terms (with respect to α) consistent with the expanded constitutive equation identified. This analysis shows that only Π and Δ appear in the zero-order terms and hence should be evaluated up to the second order. Furthermore, all of the remaining terms in Equation (5.29), except for Σ, appear only in second-order terms of the approximate equations of motion and only their leading zero-order terms need to be evaluated to preserve the consistency of the governing equations. The term Σ, which only appears in the higher-order terms of the expanded equations of motion, can be evaluated approximately using only the viscous terms. Therefore the final set of the extra stress components used in conjunction with the components of the equation of motion are

$$
\begin{cases}
\tau_{rr} = \tau_{rr}^{(0)} = \tau_{rr}^{v} \\[4pt]
\tau_{zz} = \tau_{zz}^{(0)} = \tau_{zz}^{v} \\[4pt]
\tau_{rz} = \tau_{rz}^{(0)} = \tau_{rz}^{v} \\[4pt]
\tau_{\theta r} = \tau_{\theta r}^{(0)} = \tau_{\theta r}^{v} + t_{\text{rel}}\left(\dfrac{\partial v_r}{\partial z}\tau_{\theta z}^{(0)} + \dfrac{\partial v_\theta}{\partial z}\tau_{rz}^{(0)}\right) \\[10pt]
\tau_{\theta z} = \tau_{\theta z}^{(0)} + t_{\text{rel}}\left[-v_r\dfrac{\partial \tau_{\theta z}^{(0)}}{\partial r} - v_z\dfrac{\partial \tau_{\theta z}^{(0)}}{\partial z} + \left(\dfrac{\partial v_\theta}{\partial r} - \dfrac{v_\theta}{r}\right)\tau_{rz}^{(0)} + \dfrac{\partial v_\theta}{\partial z}\tau_{zz}^{(0)} + \dfrac{v_r}{r}\tau_{\theta z}^{(0)}\right] \\[14pt]
\tau_{\theta\theta} = \tau_{\theta\theta}^{(0)} + 2t_{\text{rel}}\left[-v_r\dfrac{\partial \tau_{\theta\theta}^{(0)}}{\partial r} - v_z\dfrac{\partial \tau_{\theta\theta}^{(0)}}{\partial z} + \left(\dfrac{\partial v_\theta}{\partial r} - \dfrac{v_\theta}{r}\right)\tau_{\theta r}^{(0)} + \dfrac{v_r}{r}\tau_{\theta\theta}^{(0)}\right]
\end{cases}
\tag{5.30}
$$

where

$$
\tau_{\theta z}^{(0)} = \tau_{\theta z}^{v} \quad \text{and} \quad \tau_{\theta\theta}^{(0)} = \tau_{\theta\theta}^{v} + 2t_{\text{rel}}\frac{\partial v_\theta}{\partial z}\tau_{\theta z}^{(0)}
$$

5.4.2 Finite element discretization of the governing equations

The required working equations are derived by application of the following finite element schemes to the described governing model:

- Standard Galerkin procedure – to discretize the circumferential component of the equation of motion, Equation (5.23), for the calculation of v_θ.

- Continuous penalty method – to discretize the continuity and (r, z) components of the equation of motion, Equations (5.22) and (5.24), for the calculation of v_r and v_z. Pressure is computed via the variational recovery procedure (Chapter 3, Section 4).

- Petrov–Galerkin scheme – to discretize the energy Equation (5.25) for the calculation of T.

Elemental stiffness equations (i.e. the working equations) resulting from the described discretizations are in general written as

$$
[M]\{X\} = \{B\}
\tag{5.31}
$$

where, from Equation (5.23) for v_θ

$$
M_{ij} = \int_{\Omega_e} \eta\left[\left(\frac{\partial N_i}{\partial r}\frac{\partial N_j}{\partial r} + \frac{\partial N_i}{\partial z}\frac{\partial N_j}{\partial z}\right).r - \left(N_i\frac{\partial N_j}{\partial r} + \frac{\partial N_i}{\partial r}N_j - \frac{N_iN_j}{r}\right)\right]\mathrm{d}r\mathrm{d}z
$$

$$
+ \int_{\Omega_e} \rho\left[v_r\left(\frac{N_iN_j}{r} + N_i\frac{\partial N_j}{\partial r}\right) + v_zN_i\frac{\partial N_j}{\partial z}\right]r\,\mathrm{d}r\mathrm{d}z
\tag{5.32}
$$

For generalized Newtonian fluids the load vector (i.e. the right-hand side in Equation (5.31) is expressed as

$$B_i = \int_{\Gamma_e} N_i(\tau_{\theta r} n_r + \tau_{\theta z} n_z) r \, \mathrm{d}\Gamma_e \tag{5.33}$$

where n_r and n_z are the components of the unit vector normal to Γ_e. Note that for small cone angles n_r is at least one order of magnitude smaller than n_z and $\tau_{\theta z}$ is one order of magnitude bigger than $\tau_{\theta r}$.

For viscoelastic fluids this term is given as

$$B_i^a = B_i - \int_{\Omega_e} \left[\left(r \frac{\partial N_i}{\partial r} - N_i \right) \bar{\tau}_{\theta r} + \frac{\partial N_i}{\partial z} \bar{\tau}_{\theta z} r \right] \mathrm{d}r\mathrm{d}z - \int_{\Omega_e} \left(v_r \frac{\partial \tau_{\theta z}}{\partial r} + v_z \frac{\partial \tau_{\theta z}}{\partial z} \right) r \, \mathrm{d}r\mathrm{d}z$$

$$\tag{5.34}$$

where

$$\begin{cases} \bar{\tau}_{\theta r} = \tau_{\theta r} - \tau_{\theta r}^v \\ \bar{\tau}_{\theta z} = \tau_{\theta z} - \tau_{\theta z}^v \end{cases}$$

From the set of equations (5.22) and (5.24) (compact form based on the continuous penalty method) for v_r and v_z

$Mij =$

$$\begin{bmatrix} \int_{\Omega_e} \eta \left(2 \frac{\partial N_i}{\partial r} \frac{\partial N_j}{\partial r} + \frac{\partial N_i}{\partial z} \frac{\partial N_j}{\partial z} + \frac{2 N_i N_j}{r} \right) r \, \mathrm{d}r\mathrm{d}z + & \int_{\Omega_e} \eta \frac{\partial N_i}{\partial z} \frac{\partial N_j}{\partial r} r \, \mathrm{d}r\mathrm{d}z + \\[2mm] \int_{\Omega_e} \rho N_i \left(v_r \frac{\partial N_j}{\partial r} + v_z \frac{\partial N_j}{\partial z} \right) r \, \mathrm{d}r\mathrm{d}z + & \int_{\Omega_e} \lambda\eta \left(r \frac{\partial N_i}{\partial r} \frac{\partial N_j}{\partial z} + N_i \frac{\partial N_j}{\partial z} \right) \mathrm{d}r\mathrm{d}z \\[2mm] \int_{\Omega_e} \lambda\eta \left(r \frac{\partial N_i}{\partial r} \frac{\partial N_j}{\partial r} + \frac{\partial N_i}{\partial r} N_j + N_i \frac{\partial N_j}{\partial r} + \frac{N_i N_j}{r} \right) \mathrm{d}r\mathrm{d}z & \\[4mm] & \int_{\Omega_e} \eta \left(\frac{\partial N_i}{\partial r} \frac{\partial N_j}{\partial r} + 2 \frac{\partial N_i}{\partial z} \frac{\partial N_j}{\partial z} \right) r \, \mathrm{d}r\mathrm{d}z + \\[2mm] \int_{\Omega_e} \eta \frac{\partial N_i}{\partial r} \frac{\partial N_j}{\partial z} r \, \mathrm{d}r\mathrm{d}z + & \int_{\Omega_e} \rho N_i \left(v_r \frac{\partial N_j}{\partial r} + v_z \frac{\partial N_j}{\partial z} \right) r \, \mathrm{d}r\mathrm{d}z + \\[2mm] \int_{\Omega_e} \lambda\eta \left(r \frac{\partial N_i}{\partial z} \frac{\partial N_j}{\partial r} + \frac{\partial N_i}{\partial z} N_j \right) \mathrm{d}r\mathrm{d}z & \int_{\Omega_e} \lambda\eta \frac{\partial N_i}{\partial z} \frac{\partial N_j}{\partial z} r \mathrm{d}r\mathrm{d}z \end{bmatrix}$$

$$\tag{5.35}$$

where λ is the penalty parameter. The penalty terms in Equation (5.35) should be found using reduced integration. For generalized Newtonian fluids

$$
B_i = \left\{
\begin{array}{l}
\displaystyle \int_{\Omega_e} \rho v_\theta^2 N_i \; drdz + \int_{\Gamma_e} N_i (\tau_{rr} n_r + \tau_{rz} n_z) r \; d\Gamma_e \\[18pt]
\displaystyle \int_{\Omega_e} \rho g N_i \; r \; drdz + \int_{\Gamma_e} N_i (\tau_{zr} n_r + \tau_{zz} n_z) r \; d\Gamma_e
\end{array}
\right\}
\tag{5.36}
$$

For the viscoelastic fluids

$$
B_i = \left\{
\begin{array}{l}
\displaystyle \int_{\Omega_e} N_i \left[\rho v_\theta^2 - \bar{\tau}_{\theta\theta} - r \left(v_r \frac{\partial \tau_{\theta\theta}}{\partial r} + v_z \frac{\partial \tau_{\theta\theta}}{\partial z} \right) \right] drdz + \int_{\Gamma_e} N_i (\tau_{rr} n_r + \tau_{rz} n_z) r \; d\Gamma_e \\[18pt]
\displaystyle \int_{\Omega_e} \rho g N_i \; r \; drdz + \int_{\Gamma_e} N_i (\tau_{zr} n_r + \tau_{zz} n_z) r \; d\Gamma_e
\end{array}
\right\}
$$

$$\tag{5.37}$$

where

$$
\bar{\tau}_{\theta\theta} = \tau_{\theta\theta} - \tau_{\theta\theta}^v
$$

From Equation (5.25) for T

$$
\begin{aligned}
M_{ij} = \int_{\Omega_e} & k \left[\left(\frac{\partial N_i}{\partial r} \frac{\partial N_j}{\partial r} + \frac{\partial N_i}{\partial z} \frac{\partial N_j}{\partial z} \right) \right. \\[6pt]
& \left. - \left(\alpha_1 \frac{\partial N_i}{\partial r} + \alpha_2 \frac{\partial N_i}{\partial z} \right) \left(\frac{\partial^2 N_j}{\partial r^2} + \frac{1}{r} \frac{\partial N_j}{\partial r} + \frac{\partial^2 N_j}{\partial z^2} \right) \right] r \; drdz \\[6pt]
& + \int_{\Omega_e} \rho c \left(N_i + \alpha_1 \frac{\partial N_i}{\partial r} + \alpha_2 \frac{\partial N_i}{\partial z} \right) \left(v_r \frac{\partial N_j}{\partial r} + v_z \frac{\partial N_j}{\partial z} \right) r \; drdz
\end{aligned}
\tag{5.38}
$$

where α_1 and α_2 are upwinding parameters (Petera *et al.*, 1993). And

$$
B_i = \int_{\Omega_e} \left(N_i + \alpha_1 \frac{\partial N_i}{\partial r} + \alpha_2 \frac{\partial N_i}{\partial z} \right) \eta \dot{\gamma}^2 \; drdz + \int_{\Gamma_e} N_i \left(\frac{\partial T}{\partial r} n_r + \frac{\partial T}{\partial z} n_z \right) r \; dT
\tag{5.39}
$$

The derived working equations are solved using the following solution algorithm:

- Step 1 – start with $v_r = v_z = 0.0$ and $\eta = \eta_0$ constant.
- Step 2 – using appropriate working equations calculated v_θ.
- Step 3 – calculate $\dot{\gamma}$ using Equation (5.26).
- Step 4 – update the value of viscosity (η) using an appropriate rheological equation (e.g. temperature-dependent form of the Carreau model given as Equation (5.4)).
- Step 5 – using updated values of viscosity and calculated v_θ calculate v_r, v_z and p. For viscoelastic fluids also calculate the additional stress components at this step.
- Step 6 – update $\dot{\gamma}$ and η and calculate T.
- Step 7 – if the solution has converged stop otherwise go to the next step.
- Step 8 – update viscosity and elastic stresses and go to step 2.

Using the described algorithm the flow domain inside the cone-and-plate viscometer is simulated. In Figure 5.17 the predicted velocity field in the (r, z) plane (secondary flow regime) established inside a bi-conical rheometer for a non-Newtonian fluid is shown.

Figure 5.17 The predicted secondary flow field in the bi-conical viscometer

The stress field corresponding to this regime is shown in Figure 5.18. As this figure shows the measuring surface of the cone is affected by these secondary stresses and hence not all of the measured torque is spent on generation of the primary (i.e. viscometric) flow in the circumferential direction.

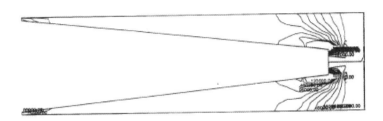

Figure 5.18 The predicted stresses arising from the secondary flow in the bi-conical viscometer

This simulation provides the quantitative measures required for evaluation of the extent of deviation from a perfect viscometric flow. Specifically, the finite element model results can be used to calculate the torque corresponding to a given set of experimentally determined material parameters as

$$\psi_{\text{fem}} = 2\pi \int\limits_{\Gamma} (\tau_{\theta r}.n_r + \tau_{\theta z}.n_z) r^2 \ \mathrm{d}\Gamma \tag{5.40}$$

where Γ represents the rotating cone surface. Through the comparison of this value with the experimentally measured torque a parameter estimation procedure can be developed which provides a strategy for the improvement of the viscometry results in this type of rheometer (Petera and Nassehi, 1995).

5.5   MODELS BASED ON THIN LAYER APPROXIMATION

There are many instances in polymer-forming processes where the flow is confined to a thin layer between relatively large surfaces. For example, in calendering, injection moulding and film blowing the flow geometry is, in general, viewed as a slowly varying thin film. The common approach adopted in the modelling of these flow systems is the utilization of the 'lubrication approximation' or its generalizations. The original lubrication approximation, proposed by Reynolds, is based on the following assumptions:

- The flow regime is steady creeping (i.e. inertia less), isothermal and dominated by viscous shear forces.

- The fluid is incompressible and Newtonian.

- One of the dimensions of the flow domain is very small in comparison to the other two dimensions.

- The small dimension (e.g. height) varies very slowly with respect to the geometrical variables in the other two directions (i.e. $\partial h/\partial x$, $\partial h/\partial y \ll 1$).

- There is no slip at the solid surfaces.

Let us consider the flow in a narrow gap between two large flat plates, as shown in Figure 5.19, where L is a characteristic length in the x and y directions and h is the characteristic gap height so that $h \ll L$. It is reasonable to assume that in this flow field $v_z \ll v_x$, v_y. Therefore for an incompressible Newtonian fluid with a constant viscosity of μ, components of the equation of motion are reduced (Middleman, 1977), as

$$\begin{cases} \dfrac{\partial p}{\partial x} = \mu \dfrac{\partial^2 v_x}{\partial z^2} \\[2mm] \dfrac{\partial p}{\partial y} = \mu \dfrac{\partial^2 v_y}{\partial z^2} \\[2mm] \dfrac{\partial p}{\partial z} = 0 \end{cases} \qquad (5.41)$$

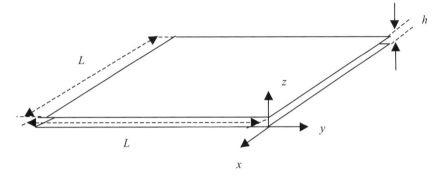

Figure 5.19 Schematic representation of a thin-layer domain between flat surfaces

According to equation set (5.41) the pressure in this flow domain is a function of x and y only and the pressure gradient normal to the parallel walls is zero. The described approximations are equivalent to the assumption of a fully developed flow between two large plates, narrowly separated with a local gap height of $h = h(x, y)$. The selection of appropriate boundary conditions for equation set (5.41) depends on the type of flow regime within the domain. For example, in the standard lubrication approximation model, the established flow in the narrow gap is considered to be a Couette regime generated by the steady motion of the top plate (Lee and Castro, 1989). In polymer processing, depending on the operating conditions, the established flow in a narrow gap may be a shear (i.e. Couette) or a pressure driven (i.e. Poiseuille) regime or a combination of both of these mechanisms. As an example, we consider a Poiseuille-type flow, characteristic of injection moulding, within the narrow gap shown in Figure 5.19. The boundary conditions corresponding to this flow are expressed as

$$\begin{cases} v_x = 0 \\ v_y = 0 \end{cases} \quad \text{at } z = 0 \text{ and } z = h \qquad (5.42)$$

Pressure gradients in the x and y directions do not depend on z, hence the integration of the components of the equation of motion, subject to conditions (5.42), yields

$$\begin{cases} v_x(z) = \dfrac{1}{2\mu}\left(\dfrac{\partial p}{\partial x}\right)(z-h)z \\[2mm] v_y(z) = \dfrac{1}{2\mu}\left(\dfrac{\partial p}{\partial y}\right)(z-h)z \end{cases} \tag{5.43}$$

Integration of the velocity components, given by Equation (5.43), with respect to gap-wise direction (i.e. z) yields the volumetric flow rates per unit width in the x and y directions as

$$\begin{cases} q_x = \displaystyle\int_0^h v_x \, dz = -\dfrac{h^3}{12\mu}\left(\dfrac{\partial p}{\partial x}\right) \\[4mm] q_y = \displaystyle\int_0^h v_y \, dz = -\dfrac{h^3}{12\mu}\left(\dfrac{\partial p}{\partial y}\right) \end{cases} \tag{5.44}$$

We now obtain the integral of the continuity equation for incompressible fluids with respect to the local gap height in this flow domain

$$\int_0^{h(x,y)} \left(\dfrac{\partial v_x}{\partial x} + \dfrac{\partial v_y}{\partial y} + \dfrac{\partial v_z}{\partial z}\right) dz = 0 \tag{5.45}$$

or

$$\int_0^h \dfrac{\partial v_x}{\partial x}\, dz + \int_0^h \dfrac{\partial v_y}{\partial y}\, dy + \int_0^h \dfrac{\partial v_z}{\partial z}\, dz = 0 \tag{5.46}$$

The last term in Equation (5.46) vanishes because $v_z = 0$ at $z = 0$ and $z = h$. Therefore

$$\dfrac{\partial}{\partial x}\int_0^h v_x \, dz + \dfrac{\partial}{\partial y}\int_0^h v_y \, dz = 0 \tag{5.47}$$

Substitution from Equation (5.44) for the integrals in Equation (5.47) gives

$$-\dfrac{1}{12\mu}\dfrac{\partial}{\partial x}\left(h^3\dfrac{\partial p}{\partial x}\right) - \dfrac{1}{12\mu}\dfrac{\partial}{\partial y}\left(h^3\dfrac{\partial p}{\partial y}\right) = 0 \tag{5.48}$$

or

$$\frac{\partial}{\partial x}\left(h^3\frac{\partial p}{\partial x}\right) + \frac{\partial}{\partial y}\left(h^3\frac{\partial p}{\partial y}\right) = 0 \tag{5.49}$$

Equation (5.49) derived for isothermal Newtonian flow in thin cavities, is called the 'pressure potential' or Hele-Shaw equation. Analogous equations in terms of pressure gradients can be obtained using other types of boundary conditions in the integration of components of the equation of motion given as Equation (5.41). The lubrication approximation approach has also been generalized to obtain solutions for non-isothermal generalized Newtonian flow in thin layers. The generalized Hele-Shaw equation for non-isothermal generalized Newtonian fluids is used extensively to model narrow gap flow regimes in injection and compression moulding (Hieber and Shen, 1980; Lee *et al.*, 1984). Other generalized equations, derived on the basis of the lubrication approximation, are used to model laminar flow in calendering, coating and other processes where the domain geometry allows utilization of this approach (Soh and Chang, 1986; Hannart and Hopfinger, 1989).

The generalized Hele-Shaw equation is expressed as

$$\frac{\partial}{\partial x}\left(S\frac{\partial p}{\partial x}\right) + \frac{\partial}{\partial y}\left(S\frac{\partial p}{\partial y}\right) = Q_0 \tag{5.50}$$

where $p(x, y)$ is pressure, Q_0 is a source term and S is called flow conductivity which for a non-elastic generalized Newtonian fluid is given (Schlichting, 1968) as

$$S = \int_0^{\frac{h}{2}} \frac{z^2}{\eta} \, \mathrm{d}z \tag{5.51}$$

where η is the apparent viscosity of a generalized Newtonian fluid. For isothermal flow of a power law fluid the flow conductivity S can be found analytically. For the non-isothermal case and when η is temperature dependent, Equation (5.50) should be solved in conjunction with the energy equation yielding the gap-wise temperature distribution, and S has to be obtained by numerical integration. To demonstrate application of the described approximation in polymer processing, in the following section modelling of flow distribution in an extrusion die is discussed.

5.5.1 Finite element modelling of flow distribution in an extrusion die

We consider a co-extrusion die consisting of an outer circular distribution channel of rectangular cross-section, connected to an extrusion slot, which is a slowly tapering narrow passage between two flat, non-parallel plates. The polymer melt is fed through an inlet into the distribution channel and flows into the tapered slot to exit from its circular outlet. The objective of simulation is to find tapering in the extrusion slot that can result in a uniform flow distribution at the die outlet. To find flow distribution the generalized Hele-Shaw equation is solved to obtain the pressure field within the domain. The pressure gradients are then used to calculate the mean velocity across the gap (or the flow rate per unit width, perpendicular to flow direction) from the following equations, derived analogous to Equations (5.43)

$$
\begin{cases}
\bar{V}_x = -2\dfrac{S}{h}\dfrac{\partial p}{\partial x} \\[2ex]
\bar{V}_y = -2\dfrac{S}{h}\dfrac{\partial p}{\partial y}
\end{cases}
\tag{5.52}
$$

Thus

$$
\begin{cases}
Q_x = h\bar{V}_x \\
Q_y = h\bar{V}_y
\end{cases}
\tag{5.53}
$$

where $\bar{V}_x$ and $\bar{V}_y$ are the components of mean velocity, Q_x and Q_y are the flow rate per unit width in the x and y directions, respectively, h is the gap height and S is defined as in Equation (5.50). To obtain the desired tapering the gap height is defined as

$$
h = a + by
\tag{5.54}
$$

where a and b are constants that should be determined through this simulation. The finite element discretization of the generalized Hele-Shaw equation (Equation (5.50)) used in this example is based on the standard Galerkin scheme. The velocity field corresponding to the simulated pressure distribution at each step is found by application of the variational recovery method (see Chapter 3, Section 1.4) to equation set (5.52). The developed solution algorithm is as follows:

- Step 1 – assume a Newtonian flow and obtain the nodal pressures.

- Step 2 – use the calculated pressures and find the velocity components by variational recovery.

- Step 3 – using the calculated velocity field, find the shear rate and update viscosity using the power law model.

- Step 4 – calculate flow conductivity, S, using updated viscosity and repeat the solution until convergence.

The described algorithm may not yield a converged solution; in particular for values of power law index less than 0.5. To ensure convergence, in the iteration cycle $(n + 1)$ for updating of the nodal pressures, an initial value found by interpolating the values of p at cycles n and $(n-1)$ should be used. Alternatively, the geometric mean values of the pressure gradients are found at each step and used in the iteration cycle to ensure convergence (Sander, 1994).

Boundary conditions for the solution of Equation (5.50) are zero pressure on the exit and zero normal pressure gradient at the inlet. The exit condition can be imposed directly; to impose the inlet conditions a layer of virtual elements is attached to the outer edges of the elements at the inlet section. A suitable value for the source term in Equation (5.50) is chosen in these elements. The source term in all other elements is set to zero. All of the boundary line integrals, appearing after the application of Green's theorem to the discretized model equation, are also set to zero. The described boundary conditions ensure, respectively, that melt leaving the extrusion slot at the exit is flowing in the radial direction, and there is no fluid flow from the outer edges of the virtual elements.

A series of simulations has been carried out, varying the parameters a and b in Equation (5.54), this is equivalent to altering the extrusion slot height and taper. The results of successive simulations are compared until a die geometry that yields uniform exit flow is obtained. In Figures 5.20a and 5.20b the exit flow uniformity, indicated by the length of the predicted velocity vectors, for sets of ($a = 0.0011$; $b = -0.0005$) and ($a = 1.3$; $b = -0.001$), respectively, is shown. As shown in Figure 5.20b, use of the second set of parameters generates an effectively uniform exit flow. This set has been found by trial and error after 10 runs (Nassehi and Pittman, 1989).

5.5.2 Generalization of the Hele-Shaw approach to flow in thin curved layers

One of the main restrictions preventing wider application of the thin-layer approach, described previously, to industrial polymer flow processes is that these models can only be used for narrow gaps between flat surfaces. Therefore generalization of the method to thin-layer flow between curved surfaces, which removes this limitation, can significantly enhance its applicability to realistic problems. To obtain such a generalization an asymptotic expansion scheme corresponding to the geometry of the thin-layer domains defined in a curvilinear coordinate system is used (Pearson and Petrie, 1970a,b; Pearson, 1985). Let us consider a thin layer between two curved non-parallel surfaces as shown in Figure 5.21: The ratio of characteristic gap height H to lateral dimensions L is $\varepsilon = (h/L) \to 0$. In the general orthogonal curvilinear coordinate system (x^i) defined for the thin domain the governing equations of the steady, creeping flow of a power law fluid are written as

continuity

$$\sum_{i=1}^{3} \frac{\partial}{\partial x^i} \left[\frac{h_1 h_2 h_3}{h_i} v(i) \right] = 0 \qquad (5.55)$$

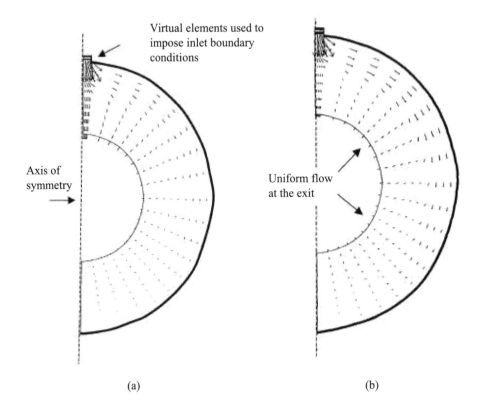

(a)   (b)

Figure 5.20   (a) The predicted non-uniform exit flow distribution in a co-extrusion die.
(b) The predicted uniform exit flow distribution obtained after altering the
gap height and tapering in the co-extrusion die

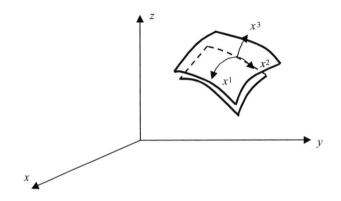

Figure 5.21   Thin layer between curved surfaces and the general curvilinear coordinate
system

motion

$$-\frac{1}{h_i}\frac{\partial p}{\partial x^i} + \sum_{j=1}^{3} \tau(ij,j) = 0 \tag{5.56}$$

where in Equations (5.55) and (5.56), h_i are the scale factors which represent the ratio of an infinitesimal displacement of a point corresponding to incremental changes of its coordinates (Aris, 1989), $v(i)$ are the physical components of the velocity and p is the pressure. In the equation of motion $\tau(ij,j)$ is expressed in terms of the components of the stress tensor as

$$\tau(ij,j) = \frac{h_i}{h_1 h_2 h_3}\frac{\partial}{\partial x^i}\left[\frac{h_1 h_2 h_3}{h_i h_j}\tau(ij)\right] + \sum_{k=1}^{3}\frac{h_i}{h_j h_k}\left\{ {}_{j}^{i}{}_{k}\right\}\tau(jk) \tag{5.57}$$

The Christoffel symbols (Spiegel, 1974), appearing in the second term on the right-hand side of Equation (5.57), are defined as

$$\left\{ {}_{l}^{l}{}_{l}\right\} = \frac{1}{h_l}\frac{\partial h_l}{\partial x^l}, \quad \left\{ {}_{l}^{l}{}_{r}\right\} = \left\{ {}_{r}^{l}{}_{l}\right\} = \frac{1}{h_l}\frac{\partial h_l}{\partial x^r}, \left\{ {}_{r}^{l}{}_{r}\right\} = -\frac{h_r}{h_l^2}\frac{\partial h_r}{\partial x^l}, \left\{ {}_{r}^{l}{}_{s}\right\} = 0$$

Here l, r and s are unequal integers in the set $\{1, 2, 3\}$. As already mentioned, in the thin-layer approach the fluid is assumed to be non-elastic and hence the stress tensor here is given in terms of the rate of deformation tensor as: $\tau(ij) = \eta D(ij)$, where, in the present analysis, viscosity η is defined using the power law equation. The model equations are non-dimensionalized using

$$\left\{\begin{array}{l} h_1 = L\hat{h}_1, \quad h_2 = L\hat{h}_2, \quad h_3 = \varepsilon L\hat{h}_3 \\ v_1 = U\hat{v}_1, \quad v_2 = U\hat{h}_2, \quad v_3 = \varepsilon U\hat{h}_3 \end{array}\right. \quad \text{and} \quad \left\{\begin{array}{l} \hat{\eta} = \eta_0 \left(\dfrac{\varepsilon L}{U}\right)^{n-1}\dot{\gamma}^{n-1} \\ p = \left(\dfrac{\eta_0 U}{\varepsilon^2 L}\right)\hat{p} \end{array}\right.$$

Non-dimensionalization of the stress is achieved via the components of the rate of deformation tensor which depend on the defined non-dimensional velocity and length variables. The selected scaling for the pressure is such that the pressure gradient balances the viscous shear stress. After substitution of the non-dimensional variables into the equation of continuity it can be divided through by $(\varepsilon L^2 U)$. Note that in the following for simplicity of writing the broken over bar on the non-dimensional variables is dropped.

Asymptotic expansion scheme

Let S_0 be a surface located at mid-channel between two smooth surfaces separated by a narrow gap. The curvilinear coordinate system, corresponding to this geometry, is constructed as (x^1, x^2) forming a set of orthogonal lines on S_0 whilst x^3 is perpendicular to this surface. The length along the x^3 coordinate line from a point on S_0 to the top and the bottom surfaces of the channel is equal and is half the local channel height.

Let H and L be two characteristic lengths associated with the channel height and the lateral dimensions of the flow domain, respectively. To obtain a uniformly valid approximation for the flow equations, in the limit of small channel thickness, the ratio of characteristic height to lateral dimensions is defined as $\varepsilon = (H/L) \rightarrow 0$. Coordinate scale factors h_i, as well as dynamic variables are represented by a power series in ε. It is expected that the scale factor h_3, in the direction normal to the layer, is $O(\varepsilon)$ while h_1 and h_2, are $O(L)$. It is also anticipated that the leading terms in the expansion of h_i are independent of the coordinate x^3. Similarly, the physical velocity components, v_1 and v_2, are $O(U)$, where U is a characteristic layer wise velocity, while v_3, the component perpendicular to the layer, is $O(\varepsilon U)$. Therefore we have

$$\begin{cases} h_1 = h_{10}(x^1, x^2) + \varepsilon h_{11}(x^1, x^2, x^3) + \varepsilon^2 h_{12}(x^1, x^2, x^3) + \ldots \\ h_2 = h_{20}(x^1, x^2) + \varepsilon h_{21}(x^1, x^2, x^3) + \varepsilon^2 h_{22}(x^1, x^2, x^3) + \ldots \\ h_3 = \qquad\qquad \varepsilon h_{31}(x^1, x^2, x^3) + \varepsilon^2 h_{32}(x^1, x^2, x^3) + \ldots \end{cases} \tag{5.58}$$

and

$$\begin{cases} U^{-1}v_1 = v_{10}(x^1, x^2) + \varepsilon v_{11}(x^1, x^2, x^3) + \varepsilon^2 v_{12}(x^1, x^2, x^3) + \ldots \\ U^{-1}v_2 = v_{20}(x^1, x^2) + \varepsilon v_{21}(x^1, x^2, x^3) + \varepsilon^2 v_{22}(x^1, x^2, x^3) + \ldots \\ U^{-1}v_3 = \qquad\qquad \varepsilon v_{31}(x^1, x^2, x^3) + \varepsilon^2 v_{32}(x^1, x^2, x^3) + \ldots \end{cases} \tag{5.59}$$

and

$$(H^2/\eta_0 UL)p = p_0 + \varepsilon p_1 + \varepsilon^2 p_2 + \ldots \tag{5.60}$$

Following the procedure for obtaining the required equations corresponding to the leading terms in the asymptotic expansions, series (5.58) to (5.60) are substituted into the governing equations. This operation is straightforward but rather lengthy, as an example, the leading terms corresponding to the components of the shear stress are derived as

$$\tau(21) = \tau(12) = \frac{\eta_0}{h_3^{n-1}}\eta\frac{U}{L}\left(\frac{1}{h_1}\frac{\partial v_2}{\partial x^1} + \frac{1}{h_2}\frac{\partial v_1}{\partial x^2} - \frac{v_2}{h_1 h_2}\frac{\partial h_2}{\partial x^1} - \frac{v_1}{h_1 h_2}\frac{\partial h_1}{\partial x^2}\right)$$

After substitution of the leading terms of the expanded variables into the model equations and equating coefficients of equal powers of ε from their sides, they are divided by common factors to obtain the following set:
Continuity equation corresponding to the first-order terms

$$\frac{\partial}{\partial x^1} (v_1 h_2 h_3) + \frac{\partial}{\partial x^2} (h_1 v_2 h_3) + \frac{\partial}{\partial x^3} (h_1 h_2 v_3) = 0 \qquad (5.61)$$

Components of the equation of motion

$$\begin{cases} \dfrac{1}{h_1} \dfrac{\partial p}{\partial x^1} = \dfrac{1}{h_3^{n+1}} \dfrac{\partial}{\partial x^3} \left(\eta \dfrac{\partial v_1}{\partial x^3} \right) \\[3mm] \dfrac{1}{h_2} \dfrac{\partial p}{\partial x^2} = \dfrac{1}{h_3^{n+1}} \dfrac{\partial}{\partial x^3} \left(\eta \dfrac{\partial v_2}{\partial x^3} \right) \\[3mm] \dfrac{1}{h_3} \dfrac{\partial p}{\partial x^3} = 0 \end{cases} \qquad (5.62)$$

Equations (5.61) and (5.62) can be used to derive a 'pressure potential' equation applicable to thin-layer flow between curved surfaces using the following procedure. In a thin-layer flow, the following velocity boundary conditions are prescribed:

$$\begin{cases} \text{Lower surface, } x^3 = x_L^3 \text{ (solid wall)} \rightarrow v_1 = v_2 = v_3 = 0 \\[3mm] \text{Middle surface, } x^3 = 0 \text{ (surface of symmetry)} \rightarrow \dfrac{\partial v_1}{\partial x^3} = \dfrac{\partial v_2}{\partial x^3} = 0, \; v_3 = 0 \\[3mm] \text{Upper surface, } x^3 = x_U^3 \text{ (solid wall)} \rightarrow v_1 = v_2 = v_3 = 0 \end{cases}$$

Integration of the first-order continuity equation between the limits of the thin layer gives

$$\int_{x_L^3}^{x_U^3} \left[\frac{\partial}{\partial x^1} (v_1 h_2 h_3) + \frac{\partial}{\partial x^2} (h_1 v_2 h_3) + \frac{\partial}{\partial x^3} (h_1 h_2 v_3) \right] dx^3 = 0 \qquad (5.63)$$

First-order terms of h_i are independent of x^3, hence

$$\frac{\partial}{\partial x^1} \left(h_2 h_3 \int_{x_L^3}^{x_U^3} v_1 \, dx^3 \right) + \frac{\partial}{\partial x^2} \left(h_1 h_3 \int_{x_L^3}^{x_U^3} v_2 \, dx^3 \right) + [h_1 h_2 v_3]_{x_L^3}^{x_U^3} = 0 \qquad (5.64)$$

After the imposition of no-slip wall boundary conditions the last term in Equation (5.64) vanishes. Therefore

$$\frac{\partial}{\partial x^1}\left(h_2 h_3 \int_{x_L^3}^{x_U^3} v_1 \, dx^3\right) + \frac{\partial}{\partial x^2}\left(h_1 h_3 \int_{x_L^3}^{x_U^3} v_2 \, dx^3\right) = 0 \tag{5.65}$$

Considering that the pressure is independent of x^3, integration of the x^1 and x^2 components of the first-order equation of motion from 0 to x^3 gives

$$\begin{cases} \left[\eta \dfrac{\partial v_1}{\partial x^3}\right]_0^{x^3} = \left(\dfrac{h_3^{n+1}}{h_1}\dfrac{\partial p}{\partial x^1}\right)x^3 \\[4mm] \left[\eta \dfrac{\partial v_2}{\partial x^3}\right]_0^{x^3} = \left(\dfrac{h_3^{n+1}}{h_2}\dfrac{\partial p}{\partial x^2}\right)x^3 \end{cases} \tag{5.66}$$

After the imposition of the boundary conditions at $x^3 = 0$ we have

$$\begin{cases} \dfrac{\partial v_1}{\partial x^3} = \dfrac{1}{\eta}\left(\dfrac{h_3^{n+1}}{h_1}\dfrac{\partial p}{\partial x^1}\right)x^3 \\[4mm] \dfrac{\partial v_2}{\partial x^3} = \dfrac{1}{\eta}\left(\dfrac{h_3^{n+1}}{h_2}\dfrac{\partial p}{\partial x^2}\right)x^3 \end{cases} \tag{5.67}$$

Second integration of components of the equation of motion yields

$$\begin{cases} \displaystyle\int_{x^3}^{x_U^3} \dfrac{\partial v_1}{\partial x^3}dx^3 = \dfrac{h_3^{n+1}}{h_1}\dfrac{\partial p}{\partial x^1}\displaystyle\int_{x^3}^{x_U^3} \dfrac{x^3\,dx^3}{\eta} \\[4mm] \displaystyle\int_{x^3}^{x_U^3} \dfrac{\partial v_2}{\partial x^3}dx^3 = \dfrac{h_3^{n+1}}{h_2}\dfrac{\partial p}{\partial x^2}\displaystyle\int_{x^3}^{x_U^3} \dfrac{x^3\,dx^3}{\eta} \end{cases} \tag{5.68}$$

and

$$\begin{cases} [v_1]_{x^3}^{x_U^3} = \dfrac{h_3^{n+1}}{h_1}\dfrac{\partial p}{\partial x^1}\displaystyle\int_{x^3}^{x_U^3} \dfrac{x^3\,dx^3}{\eta} \\[4mm] [v_2]_{x^3}^{x_U^3} = \dfrac{h_3^{n+1}}{h_2}\dfrac{\partial p}{\partial x^2}\displaystyle\int_{x^3}^{x_U^3} \dfrac{x^3\,dx^3}{\eta} \end{cases} \tag{5.69}$$

Imposition of the boundary conditions at $x^3 = x_U^3$ gives

$$
\begin{cases}
v_1 = h_3^{n+1} \Lambda_1 \int\limits_{x^3}^{x_U^3} \dfrac{x^3 \, dx^3}{\eta} \\[3em]
v_2 = h_3^{n+1} \Lambda_2 \int\limits_{x^3}^{x_U^3} \dfrac{x^3 \, dx^3}{\eta}
\end{cases}
\tag{5.70}
$$

where

$$
\Lambda_1 = -\frac{1}{h_1} \frac{\partial p}{\partial x^1}, \quad \Lambda_2 = -\frac{1}{h_2} \frac{\partial p}{\partial x^2}
$$

Assuming symmetry of v_1 and v_2 with respect to mid-surface at ($x^3 = 0$), the velocity components, given by Equation (5.70), are integrated to obtain flow rates in the lateral directions within the limits of the thin layer as

$$
\begin{cases}
\displaystyle\int\limits_{x_L^3}^{x_U^3} v_1 \, dx^3 = 2 \int\limits_{0}^{x_U^3} \left(h_3^{n+1} \Lambda_1 \int\limits_{x^3}^{x_U^3} \frac{x^3 \, dx^3}{\eta} \right) dx^3 \\[3em]
\displaystyle\int\limits_{x_L^3}^{x_U^3} v_2 \, dx^3 = 2 \int\limits_{0}^{x_U^3} \left(h_3^{n+1} \Lambda_2 \int\limits_{x^3}^{x_U^3} \frac{x^3 \, dx^3}{\eta} \right) dx^3
\end{cases}
\tag{5.71}
$$

Let $u = \int\limits_{x^3}^{x_U^3} (x^3 \, dx^3)/\eta$, $dv = dx^3$. The dummy variable x^3 is the lower limit of u, therefore we have: $du = -(x^3 \, dx^3)/\eta$, $v = x^3$. Taking terms independent of x^3 outside of the integration signs in the first equation of set (5.71)

$$
\int\limits_{x_L^3}^{x_U^3} v_1 \, dx^3 = 2 h_3^{n+1} \Lambda_1 \left(\left[x^3 \int\limits_{x^3}^{x_U^3} \frac{x^3 \, dx^3}{\eta} \right]_{0}^{x_U^3} + \int\limits_{0}^{x_U^3} \frac{(x^3)^2 \, dx^3}{\eta} \right)
\tag{5.72}
$$

The first term on the right-hand side of Equation (5.72) is zero, and

$$
\int\limits_{x_L^3}^{x_U^3} v_1 \, dx^3 = 2 h_3^{n+1} \Lambda_1 \int\limits_{0}^{x_U^3} \frac{(x^3)^2 \, dx^3}{\eta}
\tag{5.73a}
$$

Similarly

$$
\int_{x_L^3}^{x_U^3} v_1 \, dx^3 = 2h_3^{n+1}\Lambda_2 \int_0^{x_U^3} \frac{(x^3)^2 \, dx^3}{\eta} \tag{5.73b}
$$

Substituting from Equations (5.73a) and (5.73b) into the continuity equation (5.65) yields

$$
\frac{\partial}{\partial x^1}\left(h_2 h_3 . 2h_3^{n+1}\Lambda_1 \int_0^{x_U^3} \frac{(x^3)^2 \, dx^3}{\eta} \right) + \frac{\partial}{\partial x^2}\left(h_1 h_3 . 2h_3^{n+1}\Lambda_2 \int_0^{x_U^3} \frac{(x^3)^2 \, dx^3}{\eta} \right) = 0 \tag{5.74}
$$

After the substitution for Λ_1 and Λ_2 into Equation (5.74) the pressure potential equation corresponding to creeping flow of a power law fluid in a thin curved layer is derived as

$$
\frac{\partial}{\partial x^1}\left(\psi_1 \frac{\partial p}{\partial x^1} \right) + \frac{\partial}{\partial x^2}\left(\psi_2 \frac{\partial p}{\partial x^2} \right) = 0 \tag{5.75}
$$

where the flow conductivity coefficients are defined as

$$
\begin{cases}
\psi_1 = \dfrac{h_2 h_3^{n+2}}{h_1} \displaystyle\int_0^{x_U^3} \frac{(x^3)^2 \, dx^3}{\eta} \\[4mm]
\psi_2 = \dfrac{h_1 h_3^{n+2}}{h_2} \displaystyle\int_0^{x_U^3} \frac{(x^3)^2 \, dx^3}{\eta}
\end{cases} \tag{5.76}
$$

The scale factors given in the above expressions depend on the curvilinear coordinate system adopted to model a thin-layer flow. For example, the scale factors for a cylindrical coordinate system of (r, θ, z) are $h_r = 1$, $h_\theta = r$ and $h_z = 1$, and for a spherical coordinate system of (R, θ, ϕ) they are $h_R = 1$, $h_\theta = R$ and $h_\phi = R \sin \theta$ (see Appendix and Spiegel, 1974).

The comparison of flow conductivity coefficients obtained from Equation (5.76) with their counterparts, found assuming flat boundary surfaces in a thin-layer flow, provides a quantitative estimate for the error involved in ignoring the curvature of the layer. For highly viscous flows, the derived pressure potential equation should be solved in conjunction with an energy equation, obtained using an asymptotic expansion similar to the outlined procedure. This derivation is routine and to avoid repetition is not given here.

5.6   STIFFNESS ANALYSIS OF SOLID POLYMERIC MATERIALS

The focus of discussions presented so far in this publication has been on the finite element modelling of polymers as liquids. This approach is justified considering that the majority of polymer-forming operations are associated with temperatures that are above the melting points of these materials. However, solid state processing of polymers is not uncommon, furthermore, after the processing stage most polymeric materials are used as solid products. In particular, fibre- or particulate-reinforced polymers are major new material resources increasingly used by modern industry. Therefore analysis of the mechanical behaviour of solid polymers, which provides quantitative data required for their design and manufacture, is a significant aspect of the modelling of these materials. In this section, a Galerkin finite element scheme based on the continuous penalty method for elasticity analyses of different types of polymer composites is described. To develop this scheme the mathematical similarity between the Stokes flow equations for incompressible fluids and the equations of linear elasticity is utilized.

We start with the governing equations of the Stokes flow of incompressible Newtonian fluids. Using an axisymmetric (r, z) coordinate system the components of the equation of motion are hence obtained by substituting the shear-dependent viscosity in Equations (4.11) with a constant viscosity μ, as

$$\begin{cases} -\dfrac{\partial P}{\partial r} + \mu \left\{ \dfrac{\partial}{\partial r} \left[\dfrac{1}{r} \dfrac{\partial}{\partial r} (rv_r) \right] + \dfrac{\partial^2 v_r}{\partial z^2} \right\} = 0 \\[3mm] -\dfrac{\partial P}{\partial z} + \mu \left[\dfrac{1}{r} \dfrac{\partial}{\partial r} \left(r \dfrac{\partial v_z}{\partial r} \right) + \dfrac{\partial^2 v_z}{\partial z^2} \right] = 0 \end{cases} \tag{5.77}$$

And the continuity equation which is identical to Equation (4.10)

$$\frac{\partial v_r}{\partial r} + \frac{v_r}{r} + \frac{\partial v_z}{\partial z} = 0 \tag{5.78}$$

In the continuous penalty scheme used here the penalty parameter is defined as

$$\lambda = \frac{2v\mu}{(1 - 2v)} \tag{5.79}$$

where v is Poisson's ratio which is equal to 0.5 for incompressible material. As seen in Equation (5.79) the present model cannot be applied to analyse perfectly incompressible materials.

The working equation of the scheme is identical to Equation (4.70). However, by comparison it can be shown that the working equations in this scheme are identical to their counterparts obtained by the commonly used equilibrium finite element approach for elasticity analysis, provided that μ is defined as the shear modulus and λ as the bulk modulus of the material, respectively (Hughes, 1987). Note that in this case the main field variables should be regarded as displacements instead of velocity components. This similarity provides an important flexibility to switch the model from the analysis of fluid flow to solid material deformation under applied loads. The majority of polymeric materials are processed as liquids and used as solid products. Hence, the described approach has the advantage that it can predict material behaviour in both liquid and solid states using the same computer program. A full account of utilization of the present model in the analysis of polymeric composites has been published previously (Ghassemieh and Nassehi, 2001a) and here only an illustrative application is discussed.

5.6.1   Stiffness analysis of polymer composites filled with spherical particles

Bulk mechanical properties of polymeric composites, such as their modulus, depend on the properties of their constituent materials, filler/matrix volume fraction and geometrical distribution of the filler phase inside the matrix. Classification of polymer composites is based on the shape of the reinforcing phase and hence they are grouped as particulate, continuous or short-fibre composites, each associated with distinct mechanical properties and used for a different purpose. Therefore in the modelling of each class of polymer composites a different set of criteria should be considered. However, the present model has the flexibility to be used under different conditions for a wide variety of polymeric materials (Nassehi *et al.*, 1993a,b; Ghassemieh and Nassehi, 2001b,c).

In the following micro-mechanical analysis, it is assumed that the domain of interest for a polymer composite filled with spherical particles can be represented by a unit cell as shown in Figure 5.22. When this unit cell is rotated $360°$ around the axis AD, a hemisphere embedded in a cylinder is produced. The inter-particle spacing corresponding to this geometry is equal to $2(r_1 - r_2)$. Therefore the volume fraction of the filler can be calculated from the ratio r_2/r_1. For a square or cubic array arrangement the relationship between the filler volume fraction and this ratio is given as

$$V_f = \frac{\pi}{6}\left(\frac{r_2}{r_1}\right)^3 \tag{5.80}$$

While for a hexagonal array the same relationship is expressed as

$$V_f = \frac{\pi}{3\sqrt{3}}\left(\frac{r_2}{r_1}\right)^3 \tag{5.81}$$

It should be noted that the described axisymmetric unit cells do not represent actual repetitive sections of a material but their dimensions are related to the

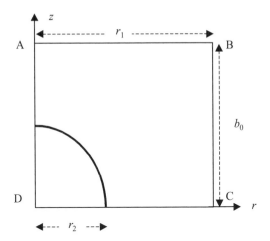

Figure 5.22 Problem domain in the micro-mechanical analysis of the particulate polymer composite

inter-particle spacing. The theoretical maximum filler volume fractions corresponding to square and hexagonal arrays are found when the ratio r_2/r_1 in Equations (5.80) and (5.81) is equal to 1. These volume fractions are hence found as 0.52 and 0.74 for square and hexagonal arrays, respectively. In practice, however, filler volume fractions in polymer composites are much lower than these theoretical values. To simulate tensile loading of the domain shown in Figure 5.22 the following boundary conditions are imposed

$$
\begin{cases}
v_z = 0 & \text{along AB } (z = 0) & (5.82\text{a}) \\
v_z = V & \text{along CD } (z = b_0) & (5.82\text{b}) \\
v_r = U & \text{along BC } (r = r_1) & (5.82\text{c}) \\
v_r = 0 & \text{along AD } (r = 0) & (5.82\text{d})
\end{cases}
$$

where V is a prescribed constant and U is determined from the condition of vanishing average lateral traction rate, defined as

$$
\int_0^{b_o} \sigma_{rr} \, dz = 0 \qquad \text{on} \qquad r = r_1 \tag{5.83}
$$

In addition to the boundary conditions (5.82a–5.82d), it is required that the displacement components should vanish on the surface of a rigid filler.

The highly constrained boundary conditions shown in Equations (5.82a) to (5.82d) can be relaxed via replacing conditions (5.82c) and (5.82d) by $\sigma_{rr} = 0$ on $r = r_1$ which is the stress-free condition along BC. Using this set of 'unconstraint' conditions the outer side wall of the cell does not remain straight and vertical under loading. The described relaxation of the boundary conditions permits consequences of deviations from highly constrained filler distribution to be investigated. Results obtained using relations (5.82a) to (5.82d) are referred to as the 'with constraint' analysis. Calculation of the field variables in the 'unconstraint' analysis is straightforward, however, to model the 'with constraint' case, the following procedure is used:

(1) The displacements and stress distribution are found imposing conditions (5.82a) to (5.82d).

     Note that the displacements are found using the working equations of the scheme; stresses are found via the variational recovery method.

(2) The field variables are recalculated replacing (5.82b) with $v_z = 0$ along CD.

(3) The first and second set of results are superimposed, therefore the final set of results are found as

$$v = v_1 + kv_2 \qquad (5.84)$$

and

$$\sigma = \sigma_1 + k\sigma_2 \qquad (5.85)$$

where k is determined such that the net force in the r direction along BC is zero, therefore

$$(F_r)_{BC} = \int_{BC} (\sigma_{r1} + k\sigma_{r2})dz = |BC|(\sigma_{r1} + k\sigma_{r2})_{BC} = 0 \qquad (5.86)$$

where

$$k = -\left(\frac{\sigma_{r1}}{\sigma_{r2}}\right)_{BC} \qquad (5.87)$$

Thus the predicted stress along AB is

$$(\sigma_z)_{AB} = (\sigma_{z1})_{AB} - \left(\frac{\sigma_{r1}}{\sigma_{r2}}\right)_{BC} (\sigma_{z2})_{AB} \qquad (5.88)$$

And because $(v_{z2})_{AB} = 0$ the displacement along this direction is expressed as

$$(v_z)_{AB} = (v_{z1})_{AB} - \left(\frac{\sigma_{r1}}{\sigma_{r2}}\right)_{BC} (v_{z2})_{AB} = (v_{z1})_{AB} \qquad (5.89)$$

Let us now consider the modulus of the composite defined as the ratio of stress over strain, i.e. $E = (\sigma_z/\varepsilon_z)$. The strain in this example is found using the specified boundary condition as

$$\varepsilon = \frac{U_z}{|BC|} = \frac{U_z}{|AB|}$$

(for $|AB| = |BC|$). The average stress along AB is also calculated as

$$\sigma_z = \frac{\displaystyle\int_A \sigma_2 \, dA}{A} \tag{5.90}$$

where A is the area of the top surface of a cylindrical unit cell in the finite element model. Using the results obtained by finite element simulations the modulus of a particulate-filled polymer composite can hence be found.

In Figure 5.23 the finite element model predictions based on 'with constraint' and 'unconstrained' boundary conditions for the modulus of a glass/epoxy resin composite for various filler volume fractions are shown.

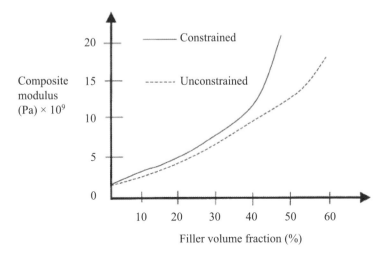

Figure 5.23   Composite modulus obtained using constrained and unconstrained boundary conditions

In Figure 5.24 the predicted direct stress distributions for a glass-filled epoxy resin under 'unconstrained' conditions for both phases are shown. The material parameters used in this calculation are: elasticity modulus and Poisson's ratio of (3.01 GPa, 0.35) for the epoxy matrix and (76.0 GPa, 0.21) for glass spheres, respectively. According to this result the position of maximum stress concentration is almost directly above the pole of the spherical particle. Therefore for a glass sphere, well bonded to the epoxy resin, the cracks in this material should start from this position and grow in the direction of direct stress. This has been experimentally confirmed using scanning electron microscopy (Ghassemieh, 1998).

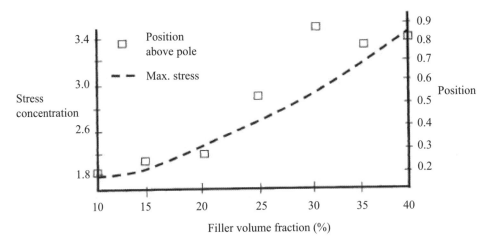

Figure 5.24   The predicted direct stress concentration at different locations within the domain

REFERENCES

Andre, J. M. *et al.*, 1998. Numerical modelling of the polymer film blowing process. *Int. J. Forming Processes* **1**, 187–210.

Aris, R., 1989. *Vectors, Tensors, and Basic Equations of Fluid Mechanics*, Dover Publications, New York.

Beaulne, M. and Mitsoulis, E. 1999. Numerical simulation of the film casting process. *Int. Poly. Process.* **XIV**, 261–275.

Bird, R. B., Armstrong, R. C. and Hassager, O., 1977. *Dynamics of Polymer Fluids, Vol. 1: Fluid Mechanics*, Wiley, New York.

Bird, R. B., Stewart, W. E. and Lightfoot, E. N., 1959. *Transport Phenomena*, Wiley, New York.

Chaturani, P. and Narasimman, S., 1990. Flow of power-law fluids in cone–plate viscometer. *Acta Mechanica* **82**, 197–211.

Clarke, J. and Freakley, P. K., 1995. Modes of dispersive mixing and filler agglomerate size distributions in rubber compounds. *Plast. Rubber Compos. Process. Appl.* **24**, 261–266.

Donea, J. and Quartapelle, L., 1992. An introduction to finite element methods for transient advection problems. *Comput. Methods Appl. Mech. Eng.* **95**, 169–203.

Ghafouri, S. N. and Freakley, P. K., A new method of flow visualisation for rubber mixing. *Pol. Test.* **13**, 171–179.

Ghassemieh, E., 1998. PhD Thesis, Department of Chemical Engineering, Loughborough University, Loughborough.

Ghassemieh, E. and Nassehi, V., 2001a. Stiffness analysis of polymeric composites using the finite element method. *Adv. Poly. Tech.* **20**, 42–57.

Ghassemieh, E. and Nassehi, V., 2001b. rediction of failure and fracture mechanisms of polymeric composites using finite element analysis. Part 1: particulate filled composites. *Poly. Compos.* **22**, 528–541.

Ghassemieh, E. and Nassehi, V., 2001c. Prediction of failure and fracture mechanisms of polymeric composites using finite element analysis. Part 2: fiber reinforced composites. *Poly. Compos.* **22**, 542–554.

Ghoreishy, M. H. R., 1997. PhD Thesis, Department of Chemical Engineering, Loughborough University, Loughborough.

Hannart, B. and Hopfinger, E. J., 1998. Laminar flow in a rectangular diffuser near Hele-Shaw conditions – a two dimensional numerical simulation. In: Bush, A. W., Lewis, B. A. and Warren, M. D. (eds), *Flow Modelling in Industrial Processes*, ch. 9, Ellis Horwood, Chichester, pp. 110–118.

Hieber, C. A. and Shen, S. F., 1980. A finite element/finite difference simulation of the injection-moulding filling process. *J. Non-Newtonian Fluid Mech.* **7**, 1–32.

Hughes, T. J. R., 1987. *The Finite Element Method*, Prentice-Hall, Englewood Cliffs NJ.

Hou, L. and Nassehi, V., 2001. Evaluation of stress – effective flow in rubber mixing. *Nonlinear Anal.* **47**, 1809–1820.

Kaye, A., Lodge, A. S. and Vale, D. G., 1968. Determination of normal stress difference in steady shear flow. *Rheol. Acta* **7**, 368–379.

Lee, C. C. and Castro, J. M., 1989. Model simplifications. In: Tucker, C L. III (ed.), *Computer Modeling for Polymer Processing*, chapter 3, Hanser Publishers, Munich, pp. 7–112.

Lee, C. C., Folgar, F. and Tucker, C. L., 1984. Simulation of compression molding for fiber-reinforced thermosetting polymers. *J. Eng. Ind.* **106**, 114–125.

Middleman, S., 1977. *Fundamentals of Polymer Processing*, McGraw-Hill, New York.

Nassehi, V., Dhillon, J. and Mascia, L., 1993a. Finite element simulation of the micromechanics of interlayered polymer/fibre composites: a study of the interactions between the reinforcing phases. *Compos. Sci. Tech.* **47**, 349–358.

Nassehi, V., Kinsella, M. and Mascia, L., 1993b. Finite element modelling of the stress distribution in polymer composites with coated fibre interlayers. *J. Compos. Mater.* **27**, 195–214.

Nassehi, V. and Ghoreishy, M. H. R., 1997. Simulation of free surface flow in partially filled internal mixers. *Int. Poly. Process.* **XII**, 346–353.

Nassehi, V. and Ghoreishy, M. H. R., 1998. Finite element analysis of mixing in partially filled twin blade internal mixers. *Int. Polym. Process.* **XIII**, 231–238.

Nassehi, V. and Ghoreishy, M. H. R., 2001. Modelling of mixing in internal mixers with long blade tips. *Adv. Polym. Technol.* **20**, 132–145.

Nassehi, V. and Pittman, J. F. T., 1989. Finite element modelling of flow distribution in an extrusion die. In: Bush, A. W., Lewis, B. A. and Warren, M. D. (eds), *Flow Modelling in Industrial Processes*, Chapter 8, Ellis Horwood, Chichester.

Nassehi, V. *et al.*, 1998. Development of a validated, predictive mathematical model for rubber mixing. *Plast. Rubber Compos.* **26**, 103–112.

Nayfeh, A. H., 1993. *Introduction to Perturbation Techniques*, Wiley, New York.

Olagunju, D. O. and Cook, L. P., 1993. Secondary flows in cone and plate flow of an Oldroyd-B fluid. *J. Non-Newtonian Fluid Mech.* **46**, 29–47.

Pearson, J. R. A., 1985. *Mechanics of Polymer Processing*, Applied Science Publishers, Barkings, Essex, UK.

Pearson, J. R. A. and Petrie, C. J. S., 1970a. The flow of a tabular film, part 1: formal mathematical representation. *J. Fluid Mech.* **40**, 1–19.

Pearson, J. R. A. and Petrie, C. J. S., 1970b. The flow of a tabular film, part 2: interpretation of the model and discussion of solutions. *J. Fluid Mech.* **42**, 609–625.

Petera, J. and Nassehi, V., 1995. Use of the finite element modelling technique for the improvement of viscometry results obtained by cone-and-plate rheometers. *J. Non-Newtonian Fluid Mech.* **58**, 1–24.

Petera, J. and Nassehi, V., 1996. Finite element modelling of free surface viscoelastic flows with particular application to rubber mixing. *Int. J. Numer. Methods Fluids* **23**, 1117–1132.

Petera, J., Nassehi, V. and Pittman, J. F. T., 1993. Petrov–Galerkin methods on isoparametric bilinear and biquadratic elements tested for a scalar convection–diffusion problem. *Int. J. Numer. Methods Heat Fluid Flow* **3**, 205–222.

Petera, J. and Pittman, J. F. T., 1994. Isoparametric Hermite elements. *Int. J. Numer. Methods Eng.* **37**, 3489–3519.

Sander, R., 1994. PhD Thesis, Chemical Engineering Department, University College of Swansea, Swansea.

Schlichting, H., 1968. *Boundary-Layer Theory*, McGraw-Hill, New York.

Soh, S. K. and Chang, C. J., 1986. Boundary conditions in the modeling of injection mold-filling of thin cavities. *Polym. Eng. Sci.* **26**, 393–399.

Spiegel, M. R., 1974. *Vector Analysis*, Schaum's outline series. McGraw-Hill, New York.

Tanner, R. I., 1985. *Engineering Rheology*, Clarendon Press, Oxford.

6

Finite Element Software – Main Components

In the finite element solution of engineering problems the main tasks of mesh generation, processing (calculations) and graphical representation of results are usually assigned to independent computer programs. These programs can either be embedded under a common shell (or interface) to enable the user to interact with all three parts in a single environment, or they can be implemented as separate sections of a software package. Development and organization of graphics programs requires expertise in areas of computer science and software design which are outside the scope of a text dealing with finite element techniques and hence are not discussed in the present chapter. Detailed explanation of mesh generation techniques and mathematical background of the available methods – although of general importance in numerical computations – are also not related to the main theme of the present book. In the following sections of this chapter therefore, after a brief description of the main aspects of mesh generation, other topics that are of central importance in the finite element modelling of polymer processes are discussed.

6.1 GENERAL CONSIDERATIONS RELATED TO FINITE ELEMENT MESH GENERATION

As discussed in the previous chapters, discretization of the solution domain into an appropriate computational mesh is the first step in the finite element simulation of field problems. Main factors in the selection of a particular mesh design for a problem are domain geometry, type of the finite elements used in the discretization, required accuracy and cost of computations. In this respect, the accuracy of computations depends on factors such as:

- consistency of the adopted mesh with the problem domain geometry,

- nature of the solution sought, and

- total number, size, aspect ratio and type of elements in the mesh.

As a general rule simulations obtained on coarse grids consisting of deformed elements of high aspect ratio are expected to have poor accuracy and should be avoided. In order to increase the accuracy of the solutions however, the mesh refinement should be based on a systematic approach which takes into account features of the physical phenomenon being analysed. Therefore it is necessary to use all available knowledge about the nature of the problem as a guide to optimize the mesh design and refinement.

6.1.1   Mesh types

Finite element solution of engineering problems may be based on a 'structured' or an 'unstructured' mesh. In a structured mesh the form of the elements and local organization of the nodes (i.e. the order of nodal connections) are independent of their position and are defined by a general rule. In an unstructured mesh the connection between neighbouring nodes varies from point to point. Therefore using a structured mesh the nodal connectivity can be implicitly defined and explicit inclusion of the connectivity in the input mesh data is not needed. Obviously this will not be possible in an unstructured mesh and nodal connectivity throughout the computational grid must be specified as part of the input data. It is important, however, to note that structured computational grids lack flexibility and hence are not suitable for engineering problems which, in general, involve complex geometries. Discretization of domains with complicated boundaries using structured grids is likely to result in badly distorted elements, thus precluding robust and accurate numerical solutions. Using an unstructured mesh, geometrical complexities can be handled in a more natural manner allowing for local adaptation, variable element concentration and preferential resolution of selected parts of the problem domain. However, because of the inherent complexity of data handling in unstructured mesh generation this approach requires special programs for the organization and recording of nodes, element edges, surfaces, etc. which involve extra memory requirement. In particular, any increase in the number of elements during mesh refinement requires rapidly rising computational efforts. A further drawback for unstructured grids is the difficulty of handling moving boundaries in a purely Lagrangian approach in the simulation of flow problems.

To resolve the problems associated with structured and unstructured grids, these fundamentally different approaches may be combined to generate mesh types which partially posses the properties of both categories. This gives rise to 'block-structured', 'overset' and 'hybrid' mesh types which under certain conditions may lead to more efficient simulations than the either class of purely structured or unstructured grids. Detailed discussions related to the properties of these classes of computational grids can be found in specialized textbooks (e.g. see Liseikin, 1999) and only brief definitions are given here.

Block-structured grids

In this approach the domain of the solution is first divided into a number of large sub-domains without leaving any gaps or overlapping. This division provides a very coarse unstructured mesh which is used as the basis for the generation of structured grids in each of its zones. The union of these local grids gives a computational grid for the entire domain, called a block-structured (or a multi-block) mesh. The flexibility gained by this approach can be used to handle complicated domains having multiply connected boundaries, problems involving heterogeneous physical phenomena and mathematical non-uniformity. Figure 6.1 shows representative examples of block-structured grids with different forms of linking or 'communication interface' between adjacent sub-regions.

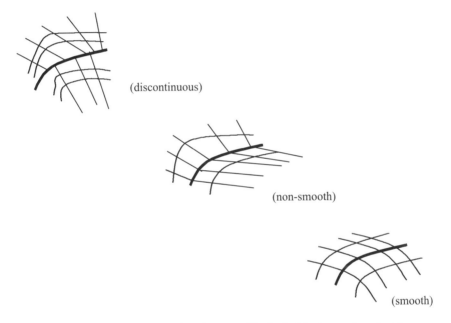

Figure 6.1   Types of interface between blocks in block-structured grids

It is important to note that finite element computations on multi-block grids involving a discontinuous interface are not straightforward and special arrangements for the transformation of nodal data across the internal boundaries are required.

Overset grids

In these girds the sub-regions or blocks are allowed to overlap and therefore the formation of a global grid is based on the assembly of individually generated structured mesh for sub-sections of the problem domain. To preserve the consistency of finite element discretizations communication between the overset regions should be based on systematic data transformation using appropriate interpolation procedures over overlapping areas of the computational grid. Figure 6.2 shows an example of this type of computational mesh.

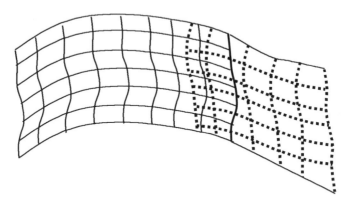

Figure 6.2   Representative fragment of an overset mesh

Hybrid grids

Hybrid grids are used for very complex geometries where combination of structured mesh segments joined by zones of unstructured mesh can provide the best approach for discretization of the problem domain. The flexibility gained by combining structured and unstructured mesh segments also provides a facility to improve accuracy of the numerical solutions for field problems of a complicated nature. Figure 6.3 shows an example of this type of computational mesh.

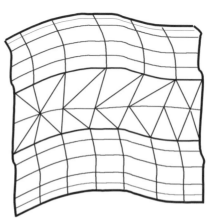

Figure 6.3   Representative fragment of a hybrid mesh

It should be emphasized at this point that the basic requirements of compatibility and consistency of finite elements used in the discretization of the domain in a field problem cannot be arbitrarily violated. Therefore, application of the previously described classes of computational grids requires systematic data transformation procedures across interfaces involving discontinuity or overlapping. For example, by the use of specially designed 'mortar elements' necessary communication between incompatible sections of a finite element grid can be established (Maday *et al.*, 1989).

6.1.2 Common methods of mesh generation

The most common approaches for the generation of structured grids are as follows:

- Algebraic methods – in these techniques calculation of grid coordinates is based on the use of interpolation formulas. The algebraic methods are fast and relatively simple but can only be used in domains with smooth and regular boundaries.

- Differential methods – in these techniques the internal grid coordinates are found via the solution of appropriate elliptic, parabolic or hyperbolic partial differential equations.

Using these procedures it is always possible to generate smooth internal divisions. Therefore they offer the advantage of preventing the extension of the exterior boundary discontinuities to the inside of the problem domain.

- Variational methods – theoretically the variational approach offers the most powerful procedure for the generation of a computational grid subject to a multiplicity of constraints such as smoothness, uniformity, adaptivity, etc. which cannot be achieved using the simpler algebraic or differential techniques. However, the development of practical variational mesh generation techniques is complicated and a universally applicable procedure is not yet available.

The most common approaches for the generation of unstructured grids are as follows:

- Octree method – this method consists of two stages. In the first stage the problem domain is covered by a Cartesian grid and during the second stage this grid is recursively subdivided. However, in this technique the computational domain boundary is effectively constructed by combining sides of grid elements and may not exactly match the prescribed problem domain boundary.

- Delaunay method – in this method the computational grid is essentially constructed by connecting a specified set of points in the problem domain. The connection of these points should, however, be based on specific rules to avoid unacceptable discretizations. To avoid breakthrough of the domain boundary it may be necessary to adjust (e.g. add) boundary points (Liseikin, 1999).

- Advancing front method – grid generation in this method starts from the boundary and is progressively moved towards the interior by the successive connection of new points appearing in front of the moving front until the entire domain is meshed into elements. At the closing stages of the procedure the advancing front should be defined in a way that it does not fold on itself. The selection of advancing step size should also be based on careful consideration and made to vary with the size of remaining unmeshed space.

In most types of unstructured grid generation a secondary smoothing is required to improve the mesh properties.

In the majority of practical finite element simulations the mesh generation is conducted in conjunction with an interactive graphics tool to allow feedback and continuous monitoring of the computational grid.

The development of more robust, accurate, flexible and versatile mesh generation methods for facilitating the application of modern computational schemes is an area of active research.

6.2   MAIN COMPONENTS OF FINITE ELEMENT PROCESSOR PROGRAMS

A typical finite element processor (sometimes called the 'number cruncher') program consists of the following blocks:

- Input and output subroutines to read and echo print data, allocate and initialize working arrays, and output the final results generally in a form that a post processor can use for graphical representations.

- Finite element library subroutines containing shape functions and their derivatives in terms of local coordinates.

- Auxiliary subroutines for handling coordinate transformation between local and global systems, quadrature, convergence checking and updating of physical parameters in non-linear calculations.

- The main subroutine for evaluation of the elemental stiffness equations and load vectors.

- Solver subroutines dealing with the assembly of elemental matrices and solution of the global set of algebraic equations.

Families of finite elements and their corresponding shape functions, schemes for derivation of the elemental stiffness equations (i.e. the working equations) and updating of non-linear physical parameters in polymer processing flow simulations have been discussed in previous chapters. However, except for a brief explanation in the worked examples in Chapter 2, any detailed discussion of the numerical solution of the global set of algebraic equations has, so far, been avoided. We now turn our attention to this important topic.

Let us first consider the assembly of elemental stiffness equations in the simple example shown in Figure 6.4.

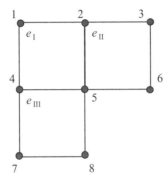

Figure 6.4   Global node numbering in a simple mesh consisting of bi-linear elements

We assume an elemental node numbering order in the clockwise direction, as shown in Figure 6.5.

Figure 6.5   Local order of node numbering

With respect to the selected elemental and global orders of node numbering the elemental stiffness equations for elements e_I, e_{II} and e_{III} in Figure 6.4 are expressed as for e_{I}

$$
\begin{bmatrix}
A_{11}^I & A_{12}^I & A_{13}^I & A_{14}^I \\
A_{21}^I & A_{22}^I & A_{23}^I & A_{24}^I \\
A_{31}^I & A_{32}^I & A_{33}^I & A_{34}^I \\
A_{41}^I & A_{42}^I & A_{43}^I & A_{44}^I
\end{bmatrix}
\begin{bmatrix}
X_4 \\ X_1 \\ X_2 \\ X_5
\end{bmatrix}
=
\begin{bmatrix}
B_1^I \\ B_2^I \\ B_3^I \\ B_4^I
\end{bmatrix}
\tag{6.1}
$$

for e_{II}

$$
\begin{bmatrix}
A_{11}^{II} & A_{12}^{II} & A_{13}^{II} & A_{14}^{II} \\
A_{21}^{II} & A_{22}^{II} & A_{23}^{II} & A_{24}^{II} \\
A_{31}^{II} & A_{32}^{II} & A_{33}^{II} & A_{34}^{II} \\
A_{41}^{II} & A_{42}^{II} & A_{43}^{II} & A_{44}^{II}
\end{bmatrix}
\begin{bmatrix}
X_5 \\ X_2 \\ X_3 \\ X_6
\end{bmatrix}
=
\begin{bmatrix}
B_1^{II} \\ B_2^{II} \\ B_3^{II} \\ B_4^{II}
\end{bmatrix}
\tag{6.2}
$$

for e_{III}

$$
\begin{bmatrix}
A_{11}^{III} & A_{12}^{III} & A_{13}^{III} & A_{14}^{III} \\
A_{21}^{III} & A_{22}^{III} & A_{23}^{III} & A_{24}^{III} \\
A_{31}^{III} & A_{32}^{III} & A_{33}^{III} & A_{34}^{III} \\
A_{41}^{III} & A_{42}^{III} & A_{43}^{III} & A_{44}^{III}
\end{bmatrix}
\begin{bmatrix}
X_7 \\ X_4 \\ X_5 \\ X_8
\end{bmatrix}
=
\begin{bmatrix}
B_1^{III} \\ B_2^{III} \\ B_3^{III} \\ B_4^{III}
\end{bmatrix}
\tag{6.3}
$$

During the assembly process the coefficients corresponding to the same degrees of freedom (i.e. the unknowns with equal indices) appearing in the elemental stiffness equations should be added together. Therefore the assembled global system for this example, arranged in ascending order of the unknowns, is expressed as

$$
\begin{bmatrix}
A_{22}^I & A_{23}^I & 0 & A_{21}^I & A_{24}^I & 0 & 0 & 0 \\
A_{32}^I & A_{33}^I + A_{22}^{II} & A_{23}^{II} & A_{31}^I & A_{34}^I + A_{21}^{II} & A_{24}^{II} & 0 & 0 \\
0 & A_{32}^{II} & A_{33}^{II} & 0 & A_{31}^{II} & A_{34}^{II} & 0 & 0 \\
A_{12}^I & A_{13}^I & 0 & A_{11}^I + A_{22}^{III} & A_{14}^I + A_{23}^{III} & 0 & A_{21}^{III} & A_{24}^{III} \\
A_{42}^I & A_{43}^I + A_{12}^{II} & A_{13}^{II} & A_{41}^I + A_{32}^{III} & A_{44}^I + A_{11}^{II} + A_{33}^{III} & A_{14}^{II} & A_{31}^{III} & A_{34}^{III} \\
0 & A_{42}^{II} & A_{43}^{II} & 0 & A_{41}^{II} & A_{44}^{II} & 0 & 0 \\
0 & 0 & 0 & A_{12}^{III} & A_{13}^{III} & 0 & A_{11}^{III} & A_{14}^{III} \\
0 & 0 & 0 & A_{42}^{III} & A_{43}^{III} & 0 & A_{41}^{III} & A_{44}^{III}
\end{bmatrix}
\begin{bmatrix}
X_1 \\ X_2 \\ X_3 \\ X_4 \\ X_5 \\ X_6 \\ X_7 \\ X_8
\end{bmatrix}
=
\begin{bmatrix}
B_2^I \\ B_3^I + B_2^{II} \\ B_3^{II} \\ B_1^I + B_2^{III} \\ B_4^I + B_1^{II} + B_3^{III} \\ B_4^{II} \\ B_1^{III} \\ B_4^{III}
\end{bmatrix}
\tag{6.4}
$$

As the number of elements in the mesh increases the sparse banded nature of the global set of equations becomes increasingly more apparent. However, as Equation (6.4) shows, unlike the one-dimensional examples given in Chapter 2, the bandwidth in the coefficient matrix in multi-dimensional problems is not constant and the main band may include zeros in its interior terms. It is of course desirable to minimize the bandwidth and, as far as possible, prevent the appearance of zeros inside the band. The order of node numbering during the mesh generation stage directly affects both of these objectives and should therefore be optimized.

In general, the imposition of boundary conditions is a part of the assembly process. A simple procedure for this is to assign a code of say 0 for an unknown degree of freedom and 1 to those that are specified as the boundary conditions. Rows and columns corresponding to the degrees of freedom marked by code 1 are eliminated from the assembled set and the other rows that contain them are modified via transfer of the product of the specified value by its corresponding coefficient to the right-hand side. The system of equations obtained after this operation is determinate and its solution yields the required results.

6.3 NUMERICAL SOLUTION OF THE GLOBAL SYSTEMS OF ALGEBRAIC EQUATIONS

Obviously selection of the most efficient elements in conjunction with the most appropriate finite element scheme is of the outmost importance in any given analysis. However, satisfaction of the criteria set by these considerations cannot guarantee or even determine the overall accuracy, cost and general efficiency of the finite element simulations, which depend more than any other factor on the algorithm used to solve the global equations. To achieve a high level of accuracy in the simulation of realistic problems usually a refined mesh consisting of hundreds or even thousands of elements is used. In comparison to time spent on the solution of the global set, the time required for evaluation and assembly of elemental stiffness equations is small. Therefore, as the number of equations in the global set grows larger by mesh refinement the computational time (and hence cost) becomes more and more dependent on the effectiveness and speed of the solver routine. The development of fast and accurate computational procedures for the solution of algebraic sets of equations has been an active area of research for many decades and a number of very efficient algorithms are now available.

As mentioned in Chapter 2, the numerical solution of the systems of algebraic equations is based on the general categories of '*direct*' or '*iterative*' procedures. In the finite element modelling of polymer processing problems the most frequently used methods are the direct methods.

Iterative solution methods are more effective for problems arising in solid mechanics and are not a common feature of the finite element modelling of polymer processes. However, under certain conditions they may provide better computer economy than direct methods. In particular, these methods have an inherent compatibility with algorithms used for parallel processing and hence are 'potentially' more suitable for three-dimensional flow modelling. In this chapter we focus on the direct methods commonly used in flow simulation models.

6.3.1   Direct solution methods

The most important direct solution algorithms used in finite element computations are based on the Gaussian elimination method.

To describe the basic concept of the Gaussian elimination method we consider
the following system of simultaneous algebraic equations

$$
\begin{bmatrix}
a_{11} & a_{12} & \cdots & a_{1n} \\
a_{21} & a_{22} & \cdots & a_{2n} \\
\cdots & \cdots & \cdots & \cdots \\
a_{n1} & a_{n2} & \cdots & a_{nn}
\end{bmatrix}
\begin{bmatrix}
x_1 \\ x_2 \\ \cdot \\ x_n
\end{bmatrix}
=
\begin{bmatrix}
b_1 \\ b_2 \\ \cdot \\ b_n
\end{bmatrix}
\tag{6.5}
$$

Let us suppose that we can convert the $n \times n$ coefficient matrix in equation
system (6.5) into an upper triangular form as

$$
\begin{bmatrix}
a^*_{11} & a^*_{12} & \cdots & a^*_{1n} \\
0 & a^*_{22} & \cdots & a^*_{2n} \\
\cdots & \cdots & \cdots & \cdots \\
0 & 0 & \cdots & a^*_{nn}
\end{bmatrix}
\begin{bmatrix}
x_1 \\ x_2 \\ \cdot \\ x_n
\end{bmatrix}
=
\begin{bmatrix}
b^*_1 \\ b^*_2 \\ \cdot \\ b^*_n
\end{bmatrix}
\tag{6.6}
$$

where all of the elements in the coefficient matrix which are below the main
diagonal are zero. The superscripts (*) in the non-zero terms of the coefficient
matrix and the right-hand side of Equation (6.6) signify the change in the values
of these components during the conversion of the original system to the upper
triangular form. It is evident that x_n can be found immediately from the last
equation in this system and the solution can hence progress by substitution of its
value in the penultimate equation to find x_{n-1} and so on.

The Gaussian elimination method provides a systematic approach for implementation of the described 'forward reduction' and 'back substitution' processes
for large systems of algebraic equations.

Pivoting

It is readily recognized that in order to generate zeros in the columns of the
coefficient matrix in an $n \times n$ system of algebraic equations multiples of successive rows should be subtracted from the rows which are above them. For
example, after multiplication of the second row in Equation (6.5) by a_{11}/a_{21} and
subtraction of the result from the first row the first component of the new
second row will be zero. However, in a large set of equations with widely
variable coefficients this operation may give rise to very large or very small
numbers causing overflow or underflow in the computer system operations
rendering the solution impossible or inaccurate. There is also the possibility of
creation of zeros in the main diagonal of the system of equations that makes the
entire set singular and hence not solvable. To avoid these problems a procedure

known as *pivoting* is used. This is described as rearranging the ordering of the equations in the set, in every step of the forward reduction, so that the coefficient of largest magnitude is located on the main diagonal. This can be achieved by interchanging rows and columns in the set. If both row and column interchanges are carried out the process is called *full pivoting*. However, this is seldom necessary as the coefficient of largest magnitude can usually be placed on the main diagonal by *partial pivoting* which only requires row interchange.

It is evident that multiplication of the sides of an equation in a system by a large number will affect the pivot selection. In particular if the scaling of the equations in full pivoting is ignored equations with larger coefficients will be positioned at rows above those having smaller coefficients. Repetition of this process during the reduction stage may lead to a situation in which the coefficients of equations located at the bottom rows are insignificant in comparison to the values of the terms at the top rows. This can become a source of unacceptable computational errors. To avoid dependence of pivoting on the scaling of the equations, the system should be normalized to make the largest coefficient in each row equal to unity (Press *et al.*, 1987).

The use of a uniform scale in partial pivoting can also significantly reduce round off errors (Gerald and Wheatley, 1984).

Gaussian elimination with partial pivoting

The solution of linear algebraic equations by this method is based on the following steps:

- Step 1 – the $n \times n$ coefficient matrix is augmented with the load vector on the right-hand side to form an $n \times (n + 1)$ matrix.

- Step 2 – interchanging rows the value of a_{11} is made to be the coefficient of largest magnitude in the first column.

- Step 3 – subtracting a_{i1}/a_{11} times the first row from the ith row, the coefficients in the first column from the second through nth rows are made equal to zero. The multiplier a_{i1}/a_{11} is stored in a_{i1}, $i = 2, \ldots, n$.

- Step 4 – steps 2 and 3 are repeated for the second through the $(n - 1)$st rows, placing the coefficient of largest magnitude on the diagonal by interchanging rows (for only rows j to n) and subtracting a_{ij}/a_{jj} times the jth row from the ith row to create zeros in all positions of the jth column below the diagonal. The multiplier a_{ij}/a_{jj} is stored in a_{ij}, $i = j + 1, \ldots, n$ (note that a_{ij} and a_{jj} used during this step are different from their initial values given in the original set of equations). At the end of this step the forward reduction processes are carried to completion and the original $n \times n$ system is converted into an upper triangular form.

- Step 5 – the last equation is solved to give $x_n = a^*_{n,n+1}/a^*_{nn}$.

- Step 6 – the remaining unknowns are found by back substitution using the following formula from the $(n-1)$st to the first equation in turn

$$x_i = \frac{1}{a^*_{ii}} \left(a^*_{i,n+1} - \sum_{j=i+1}^{n} a^*_{ij} x_j \right)$$

Number of operations in the Gaussian elimination method

To estimate the computational time required in a Gaussian elimination procedure we need to evaluate the number of arithmetic operations during the forward reduction and back substitution processes. Obviously multiplication and division take much longer time than addition and subtraction and hence the total time required for the latter operations, especially in large systems of equations, is relatively small and can be ignored. Let us consider a system of simultaneous algebraic equations, the representative calculation for forward reduction at stage is expressed as

$$a^k_{ij} = a^{k-1}_{ij} - \frac{a^{k-1}_{ik}}{a^{k-1}_{kk}} a^{k-1}_{kj} \qquad \text{where} \qquad \begin{cases} k = 1, \ldots, (n-1) \\ j = (k+1), \ldots, (n+1) \\ i = (k+1), \ldots, (n) \end{cases} \qquad (6.7)$$

The augmented coefficient matrix at this stage can be shown as

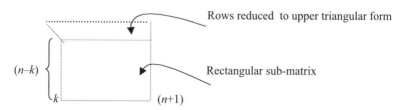

$(n-k)$ — Rows reduced to upper triangular form — Rectangular sub-matrix — k — $(n+1)$

As Equation (6.7) shows in each of $(n-k)$ rows in the rectangular sub-matrix we need to evaluate $(n-k)$ multipliers and carry out $(n-k+1)$ multiplications, therefore the total number of operations required is calculated as

$$S_t = \sum_{k=1}^{n-1} [(n-k)(n-k+1) + (n-k)] \approx$$

$$\int_{k=1}^{n} [(n-k)(n-k+1) + (n-k)] \mathrm{d}k = \tfrac{1}{3} n^3 - n \qquad (6.8)$$

For large n, $n^3 \gg n$ and $S_t = 1/3\ n^3$.

The described 'operations count' provides a guide to estimate the computational time required for reduction of a full $n \times n$ matrix to upper triangular form. However, the global set of equations in finite element analysis will always be represented by a sparse banded (it may also be symmetric) coefficient matrix. It is therefore natural to consider ways for the exclusion of zero terms from arithmetic operations during computer implementation of the Gaussian elimination method. An additional advantage of modification of the basic procedure to enable the forward reduction to be applied only to the non-zero terms is to reduce the storage (i.e. core) requirement. To take full advantage of this possibility it is important to optimize the global node numbering in the finite element mesh in a way that the maximum bandwidth of non-zero terms remains as small as possible and creation of zeros in the interior elements of the band is avoided. Efficient band solver procedures such as the *active column* or *skyline reduction method* are now available (Bathe, 1996) which provide maximum computer economy by restricting the number of operations and high-speed storage requirement.

6.4 SOLUTION ALGORITHMS BASED ON THE GAUSSIAN ELIMINATION METHOD

The most frequently used modifications of the basic Gaussian elimination method in finite element analysis are the '*LU decomposition*' and '*frontal solution*' techniques.

6.4.1 LU decomposition technique

This technique (also known as the Crout reduction or Cholesky factorization) is based on the transformation of the matrix of coefficients in a system of algebraic equations into the product of lower and upper triangular matrices as

$$
\begin{bmatrix}
a_{11} & a_{12} & \cdots & a_{1n} \\
a_{21} & a_{22} & \cdots & a_{2n} \\
\cdots & \cdots & \cdots & \cdots \\
a_{n1} & a_{n2} & \cdots & a_{nn}
\end{bmatrix}
=
\begin{bmatrix}
l_{11} & 0 & \cdots & 0 \\
l_{21} & l_{22} & \cdots & 0 \\
\cdots & \cdots & \cdots & \cdots \\
l_{n1} & l_{n2} & \cdots & l_{nn}
\end{bmatrix}
\cdot
\begin{bmatrix}
1 & u_{12} & \cdots & u_{1n} \\
0 & 1 & \cdots & u_{2n} \\
\cdots & \cdots & \cdots & \cdots \\
0 & 0 & \cdots & 1
\end{bmatrix}
\tag{6.9}
$$

Therefore $a_{11} = l_{11} \times 1 = l_{11}$, $a_{21} = l_{21}$, etc. (elements in the first column of a are the same as the elements in the first column of l); similarly multiplying rows of l by columns of u and equating the result with the corresponding element of a all of the elements of lower and upper triangular matrices are found. The general formula for obtaining elements of l and u can be expressed as

$$
\begin{cases}
l_{ij} = a_{ij} - \sum_{k=1}^{j-1} l_{ik} u_{kj}, & j \le i, i = 1, 2, \ldots, n \qquad (\text{for } j = 1, l_{i1} = a_{i1}) \\[2em]
u_{ij} = \dfrac{a_{ij} - \sum_{k=1}^{j-1} l_{ik} u_{kj}}{l_{ii}}, & i \le j, j = 2, 3, \ldots, n \qquad \left(\text{for } i = 1, u_{1j} = \dfrac{a_{1j}}{a_{11}}\right)
\end{cases}
\tag{6.10}
$$

After obtaining the described decomposition the set of equations can be readily solved. This is because all of the information required for transformation of the coefficient matrix to an upper triangular form is essentially recorded in the lower triangle. Therefore modification of the right-hand side is quite straightforward and can be achieved using the lower triangular matrix as

$$
\begin{cases}
b_1{}^* = \dfrac{b_1}{l_{11}} \\[2em]
b_i{}^* = \dfrac{b_i - \sum_{k=1}^{i-1} l_{ik} b_k{}^*}{l_{ii}}, & i = 2, 3, \ldots, n
\end{cases}
\tag{6.11}
$$

Hence the solution is found by back substitution based on

$$
\begin{cases}
x_n = b_n{}^* \\[1.5em]
x_j = b_j{}^* - \sum_{k=j+1}^{n} u_{jk} x_k, & j = n-1, n-2, \ldots, 1
\end{cases}
\tag{6.12}
$$

In some applications the diagonal elements of the upper triangular matrix are not predetermined to be unity. The formula used for the LU decomposition procedure in these applications is slightly different from those given in Equations (6.10) to (6.12), (Press *et al.*, 1987).

The LU decomposition procedure used in conjunction with partial pivoting provides a very efficient method for the solution of systems of algebraic equations. The main advantage of this approach over the basic Gaussian elimination method is that once the coefficient matrix is decomposed into a product of lower and upper triangular matrices it remains the same while the right-hand side can be changed. Therefore different solutions for a set of algebraic equations with different right-hand sides can be found rapidly. In practice this property can be used to investigate the effect of altering boundary conditions in a field problem with maximum computing economy. This property of the LU decomposition procedure can also be utilized to minimize the computational cost of iterative improvement of the accuracy of the solution of systems of linear equations (see Section 5.2).

6.4.2   Frontal solution technique

Computer implementations of band solver routines based on methods such as LU decomposition essentially depend on the 'in core' handling of the totally assembled elemental stiffness equations. This may prevent the use of small PCs (or even medium-size workstations) for simulation of realistic engineering problems, which require a relatively refined finite element mesh. The frontal solution procedure, originally developed by Irons (1970), avoids this problem by piecemeal reduction of the total matrix (or non-zero band) in a Gaussian elimination procedure. The original routine handled the solution of symmetric positive-definite matrix equations, however, in many problems (especially in the finite element simulation of flow processes) the equations to be solved are non-symmetric. Therefore, in flow modelling a non-symmetric version of the original algorithm, developed by Hood (1976), is usually used. The basic concept of the frontal solution strategy is as follows:

A work array of limited size (say $d \times d$, where d is called the front width) is selected as the pre-assigned core area for the assembly, pivoting and reduction of elemental stiffness equations. Using a loop elemental stiffness matrices are assembled until the work array is filled. In this limited area of the total matrix pivoting is implemented and forward reduction is carried out. After elimination of a sufficient number of coefficients in the work array further assembly becomes possible and the cycle can be repeated. The progress of assembly and elimination and position of the front in a finite element mesh is shown, schematically, in Figure 6.6. The active front shown in the figure implies that, at this stage of the process, coefficients of elemental stiffness equations for elements 1 to 4 corresponding to variables which are not on the front have already been fully assembled and reduced.

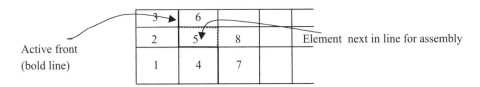

Figure 6.6   Frontal solution scheme

Frontal solution requires very intricate bookkeeping for tracking coefficients and making sure that all of the stiffness equations have been assembled and fully reduced. The process time requirement in frontal solvers is hence larger than a straightforward band solver for equal size problems.

Another consequence of using this strategy is that, unlike band solver routines, global node numbering in frontal solvers may be done in a completely arbitrary manner. However, better computer economy is achieved if an element numbering which minimizes front width is used. In general, manipulation of element numbering in a global domain is much simpler than nodes, consequently mesh design optimization in programs using frontal solvers is simpler than those based on band solver routines.

6.5   COMPUTATIONAL ERRORS

As mentioned earlier, overall accuracy of finite element computations is directly determined by the accuracy of the method employed to obtain the numerical solution of the global system of algebraic equations. In practical simulations, therefore, computational errors which are liable to affect the solution of global stiffness equations should be carefully analysed.

6.5.1   Round-off error

All computer systems operate under a predetermined floating-point word length which automatically imposes the rounding-off of digits beyond a given limit in all calculations. It is of course possible to use double-precision arithmetic to increase this range but the restriction cannot be removed completely. Therefore through accumulation of round-off errors, especially in simulations involving large numbers of calculations, serious computational problems may arise. In the solution of simultaneous algebraic equations, severity of pathological situations related to round-off errors depends on the conditioning of the coefficient matrix in the system. Conditioning of a matrix depends on:

- how small (i.e. near zero) is its smallest eigenvalue and, more importantly
- how large is the ratio of the largest to the smallest eigenvalue.

The degree of conditioning of a matrix is determined by the '*condition number*' defined as (Fox and Mayers, 1977)

$$\text{cond}\,(A) = \frac{\lambda_n}{\lambda_1} \tag{6.13}$$

where λ_n and λ_1 are the largest and smallest eigenvalues of A, respectively. In practice the condition number is found approximately using the upper bound of λ_n as $\lambda_n^u = \|A\|$, where $\|A\|$ represents any matrix norm, and a lower bound for λ_1, using the method of inverse iteration for the calculation of eigenvectors (Bathe, 1996). Therefore

$$\text{cond}\,(A) \approx \frac{\lambda_n^u}{\lambda_1} \tag{6.14}$$

A matrix with a large condition number is commonly referred to as 'ill-conditioned' and particularly vulnerable to round-off errors. Special techniques, generally known as preconditioning (Gluob and Van Loan, 1984), are used to overcome problems associated with the numerical solution of ill-conditioned matrices.

6.5.2   Iterative improvement of the solution of systems of linear equations

Consider the solution of a set of linear algebraic equations given as

$$[A]\{x\} = \{b\} \tag{6.15}$$

Let us assume that a numerical solution for this set is found as $\{x\}+\{\delta x\}$, where $\{\delta x\}$ is an unknown error. Therefore insertion of this result into the original equation set should give a right-hand side which is different from the true $\{b\}$. Thus

$$[A]\{x + \delta x\} = \{b + \delta b\} \tag{6.16}$$

Subtraction of Equation (6.15) from Equation (6.16) gives

$$[A]\{\delta x\} = \{\delta b\} \tag{6.17}$$

Replacing for $\{\delta b\}$ from Equation (6.16) in Equation (6.17) gives

$$[A]\{\delta x\} = [A]\{x + \delta x\} - \{b\} \tag{6.18}$$

The right-hand side in Equation (6.18) is known and hence its solution yields the error $\{\delta x\}$ in the original solution. The procedure can be iterated to improve the solution step-by-step. Note that implementation of this algorithm in the context of finite element computations may be very expensive. A significant advantage of the LU decomposition technique now becomes clear, because using this technique $[A]$ can be decomposed only once and stored. Therefore in the solution of Equation (6.18) only the right-hand side needs to be calculated.

REFERENCES

Bathe, K. J., 1996. *Finite Element Procedure*, Prentice Hall, Englewood Cliff, NJ.

Fox, L. and Mayers, D. F., 1977. *Computing Methods for Scientists and Engineers*, Clarendon Press, Oxford.

Gerald, C. F. and Wheatley, P. O., 1984. *Applied Numerical Analysis*, 3rd edn, Addison-Wesley, Reading, MA.

Gluob, G. H. and Van Loan, C. F., 1984. *Matrix Computations*, Johns Hopkins University Press, Baltimore.

Hood, P., 1976. Frontal solution program for unsymmetric matrices. *Int. J. Numer. Methods Eng.* **10**, 379–399.

Irons, B. M., 1970. A frontal solution for finite element analysis. *Int. J. Numer. Methods Eng.* **2**, 5–32.

Liseikin, V. D., 1999. *Grid Generation Methods*, Springer-Verlag, Berlin.

Maday, Y., Mavripilis, C. and Petera, A. T., 1980. Nonconforming mortar element methods: application to spectral discretizations. In: Chan, T. F., Glowinski, R., Periaux, J. and Widlund, O. B. (eds), *Domain Decomposition Methods*, SIAM, Philadelphia, pp. 392–418.

Press, W. H. *et al.*, 1987. *Numerical Recipes – The Art of Scientific Computing*, Cambridge University Press, Cambridge.

7

Computer Simulations – Finite Element Program

In this chapter details of a computer program entitled PPVN.f are described. This program is based on application of the Galerkin finite element method to the solution of the governing equations of generalized Newtonian flow. Flow equations are solved by the continuous penalty scheme and streamline upwinding is applied to solve the convection-dominated energy equation. Nodal values of pressure (and stress components) are found by the variational recovery method. In the last section of this chapter the program source code is listed which includes basic subroutines required in the finite element simulation of non-isothermal incompressible generalized Newtonian flow regimes. The central concern in the development of PPVN.f has been to adopt a programming style that makes the modification or extension of the code as convenient as possible for the user. As an example, the modifications required for extension of the program to axisymmetric domains are discussed. It is shown that by comparing the working equations used in PPVN.f, with their counterparts for axisymmetric domains, necessary modifications for this extension can be readily identified and implemented. In the following sections, the solution algorithm used in PPVN.f and the function of each subroutine in the code are explained. A manual describing the structure of the input data file is also presented. Finally, a simple example of the application of the program is included which shows the simulation of non-isothermal flow of a power law fluid in a two-dimensional domain.

7.1 PROGRAM STRUCTURE AND ALGORITHM

PPVN.f is a FORTRAN program for the solution of steady state, generalized Navier–Stokes and energy equations in two-dimensional planar domains. This program uses a decoupled algorithm to solve the flow and energy equations iteratively. The steps used in this algorithm are shown in the following chart:

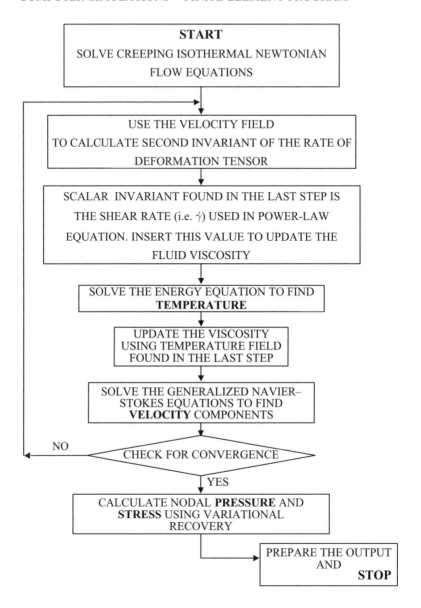

7.2 PROGRAM SPECIFICATIONS

PPVN.f consists of a main module and 24 subroutines. These subroutines and their assigned tasks are described in this section. Variables in these subroutines are all defined using 'comment' statements inserted in the listing of the program code.

MAIN: The tasks allocated to the main module are:

(i)   Defines and allocates memory spaces and initializes variable arrays used in the program.

(ii)   Opens formatted file channels for input and output and the scratch files that are required during the calculations.

(iii)   Asks the user to give the input data file name (interactively).

(iv)   Reads the input data file. The 'read' statements are either included in the main module or in the subroutines that are designed to read specific parts of the data file.

(v)   Checks the acceptability of the input data.

(vi)   Calls subroutines that prepare work arrays and specify positions in the global system of equations where the prescribed boundary conditions should be inserted.

(vii)   Starts the 'solution' loop.

(viii) Prepares the output files.

GAUSSP. Gives Gauss point coordinates and weights required in the numerical integration of the members of the elemental stiffness equations.

SHAPE. Gives the shape functions in terms of local coordinates for bi-linear or bi-quadratic quadrilateral elements.

DERIV. Calculates the inverse of the Jacobian matrix used in isoparametric transformations.

SECINV. Calculates the second invariant of the rate of deformation tensor at the integration points within the elements.

FLOW. Calculates members of the elemental stiffness matrix corresponding to the flow model.

ENERGY. Calculates members of the elemental stiffness matrix corresponding to the energy equations.

STRESS. Applies the variational recovery method to calculate nodal values of pressure and, components of the stress. A mass lumping routine is called by STRESS to diagonalize the coefficient matrix in the equations to eliminate the reduction stage of the direct solution. Prints out nodal pressure and stress components. These results should be stored in a suitable file for post-processing.

LUMPM: Diagonalizes a square matrix.

SOLVER: Assembles elemental stiffness equations into a banded global matrix, imposes boundary conditions and solves the set of banded equations using the LU decomposition method (Gerald and Wheatley, 1984). SOLVER calls the following 4 subroutines:

- ASEMB – reads and assembles the elemental stiffness equations in a banded form, as is illustrated in Figure 7.1 to minimize computer memory requirements.

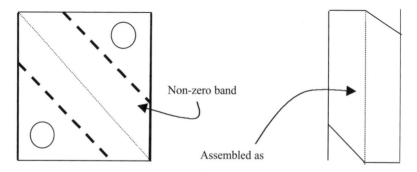

Figure 7.1   Arrangement of the global matrix in the band-solver subroutine

- BANDWD – calculates the maximum bandwidth of non-zero terms in the coefficient matrix.

- MODIFY – addressing of members of the coefficient matrix are adjusted to allocate their row and column index in the banded matrix.

- SOLVE – inserts the prescribed boundary conditions and uses an LU decomposition method to solve the assembled equations.

GETNOD: Reads and echo prints nodal coordinates; formatting should match the output generated by the pre-processor.

GETELM: Reads and echo prints element connectivity; formatting should match the output generated by the pre-processor.

GETBCD: Reads and echo prints prescribed boundary conditions, formatting should match the output generated by the pre-processor.

PUTBCV: Inserts the prescribed velocity boundary values at the allocated place in the vector of unknowns for flow equations.

PUTBCT: Inserts the prescribed temperature boundary values at the allocated place in the vector of unknowns for the energy equation.

CLEAN: Cleans used arrays at the end of each segment and prepares them to be used by the next component of the program.

SETPRM: Rearranges numbers of nodal degrees of freedom to make them compatible with the velocity components at each node. For example, in a nine-noded element allocated degree of freedom numbers for v_1 and v_2 at node n are X_n and X_{n+9}, respectively.

GETMAT: Reads and echo prints physical and rheological parameters and the penalty parameter used in the simulation.

VISCA: Calculates shear dependent viscosity using the power law model.

CONTOL: Calculates ratio of the difference of the Euclidean norm (Lapidus and Pinder, 1982) between successive iterations to the norm of the solution, as

$$\sqrt{\frac{\sum_{i=1}^{N}|X_i^{r+1} - X_i^r|^2}{\sum_{i=1}^{N}|X_i^{r+1}|^2}} \leq \varepsilon \tag{7.1}$$

where r is the number of the iteration cycle, N is the total number of degrees of freedom and ε is the convergence tolerance value. Note that criterion (7.1) is used for both velocity components and temperature in separate calculations. A converged solution is obtained when both sets of results satisfy this criterion.

OUTPUT: Prints out the computed velocity and temperature fields. For post-processing the user should store the output in suitable files

7.3 INPUT DATA FILE

Structure of the input data file required for the running of PPVN.f is as follows:

Line 1 Title

format(A40)

Line 2 Basic control variables

format(3i5, 2f10.0)

ncn	=	number of nodes per element
ngaus	=	number of full integration points
mgaus	=	number of reduced integration points
alpha	=	parameter to mark shear terms in the flow equations
beta	=	parameter to mark penalty terms in the flow equations

Line 3 Mesh data

format(3i5)

nnp	=	total number of nodes
nel	=	total number of elements
nbc	=	total number of boundary conditions

Line 4 Convergence tolerance

format(2f10.5)

tolv	=	tolerance parameter for the convergence of velocity calculations
tolt	=	tolerance parameter for the convergence of temperature calculations

Line 5 Rheological and physical parameters

format(7f10.0)

rvisc	=	consistency coefficient in the power law model
power	=	power law index
tref	=	reference temperature in temperature-dependent power law model
tbco	=	temperature dependency coefficient in the power law model
roden	=	fluid density
cp	=	specific heat capacity of the fluid
condk	=	conductivity coefficient

Line 6 Penalty parameter

format(f15.0)

rplam	=	penalty parameter

A pre-processor program should usually be used to generate the following data lines

Lines 7 to ip = 7 + nnp   nodal coordinates

$$format(i5,2e15.8)$$

cord (maxnp, ndim) = coordinates of the nodal points

Lines ip to ie = ip + nel   element connectivity

$$format(10i5)$$

node (iel, icn) = array consisting of element numbers and nodal connectivity

Lines ie to ib = ie + nbc   boundary conditions

$$format(2i5,f10.0)$$

ibc (maxbc) = numbers of nodes where a boundary condition is given
jbc (maxbc) = index to identify the prescribed degree of freedom (e.g. jbc = 1
              the first component of velocity is given etc.)
vbc (maxbc)= value of the prescribed boundary conditions

7.4 EXTENSION OF PPVN.F TO AXISYMMETRIC PROBLEMS

As already mentioned, the present code corresponds to the solution of steady-state non-isothermal Navier–Stokes equations in two-dimensional Cartesian domains by the continuous penalty method. As an example, we consider modifications required to extend the program to the solution of creeping (Stokes) non-isothermal flow in axisymmetric domains:

Step 1 To solve a Stokes flow problem by this program the inertia term in the elemental stiffness matrix should be eliminated. Multiplication of the density variable by zero enforces this conversion (this variable is identified in the program listing).

Step 2 General structure of stiffness matrices derived for the model equations of Stokes flow in (x, y) and (r, z) formulations (see Chapter 4) are compared.

Stiffness matrix corresponding to flow equations in (x, y) formulation

$$
\begin{bmatrix}
\int_{\Omega_e} \left[(\lambda + 2\eta) \frac{\partial N_i}{\partial x} \cdot \frac{\partial N_j}{\partial x} + \eta \frac{\partial N_i}{\partial y} \cdot \frac{\partial N_j}{\partial y} \right] dxdy & \int_{\Omega_e} \left(\lambda \frac{\partial N_i}{\partial x} \cdot \frac{\partial N_j}{\partial y} + \eta \frac{\partial N_i}{\partial y} \cdot \frac{\partial N_j}{\partial x} \right) dxdy \\[4mm]
\int_{\Omega_e} \left(\lambda \frac{\partial N_i}{\partial y} \cdot \frac{\partial N_j}{\partial x} + \eta \frac{\partial N_i}{\partial x} \cdot \frac{\partial N_j}{\partial y} \right) dxdy & \int_{\Omega_e} \left[(\lambda + 2\eta) \frac{\partial N_i}{\partial y} \cdot \frac{\partial N_j}{\partial y} + \eta \frac{\partial N_i}{\partial x} \cdot \frac{\partial N_j}{\partial x} \right] dxdy
\end{bmatrix}
$$

(7.2)

Stiffness matrix corresponding to flow equations in (r, z) formulation

$$
\begin{bmatrix}
\begin{aligned}
& \int_{\Omega_e} \lambda \left(\frac{N_i N_j}{r^2} + \frac{N_i}{r} \frac{\partial N_j}{\partial r} + \frac{\partial N_i}{\partial r} \frac{N_j}{r} \right. \\
& \qquad\qquad \left. + \frac{\partial N_i}{\partial r} \frac{\partial N_j}{\partial r} \right) r\, drdz \\
& + \int_{\Omega_e} \eta \left(\frac{2 N_i N_j}{r^2} + 2 \frac{\partial N_i}{\partial r} \frac{\partial N_j}{\partial r} \right. \\
& \qquad\qquad \left. + \frac{\partial N_i}{\partial z} \frac{\partial N_j}{\partial z} \right) r\, drdz
\end{aligned}
&
\begin{aligned}
& \int_{\Omega_e} \lambda \left(\frac{\partial N_i}{\partial r} \frac{\partial N_j}{\partial z} + \frac{N_i}{r} \frac{\partial N_j}{\partial z} \right) r\, drdz \\
& + \int_{\Omega_e} \eta \left(\frac{\partial N_i}{\partial z} \frac{\partial N_j}{\partial r} \right) r\, drdz
\end{aligned} \\[10mm]
\begin{aligned}
& \int_{\Omega_e} \lambda \left(\frac{\partial N_i}{\partial z} \frac{\partial N_j}{\partial r} + \frac{\partial N_i}{\partial z} \frac{N_j}{r} \right) r\, drdz + \\
& \int_{\Omega_e} \eta \left(\frac{\partial N_i}{\partial r} \frac{\partial N_j}{\partial z} \right) r\, drdz
\end{aligned}
&
\begin{aligned}
& \int_{\Omega_e} \lambda \left(\frac{\partial N_i}{\partial z} \frac{\partial N_j}{\partial z} \right) r\, drdz \\
& + \int_{\Omega_e} \eta \left(2 \frac{\partial N_i}{\partial z} \frac{\partial N_j}{\partial z} + \frac{\partial N_i}{\partial r} \frac{\partial N_j}{\partial r} \right) r\, drdz
\end{aligned}
\end{bmatrix}
$$

(7.3)

Step 3   Comparing systems (7.2) and (7.3) additional terms in the members of the stiffness matrix corresponding to the axisymmetric formulation are identified. Note that the measure of integration in these terms is $(r\, drdz)$.

Step 4   Modification of subroutine SECINV:

- Define (r) in the program and find its value at the integration points.

- Find radial component of velocity (v_r) at the integration points.

- Calculate $1/2(v_r^2/r^2)$ and add this value to the previously calculated value of the shear rate.

Step 5   Modification of subroutine FLOW:

- Define (r) in the program and find its value at the integration points. The measure of integration should be multiplied by this factor.

- After evaluation of the terms of the stiffness matrix modify them according to the additional terms shown in system (7.3).

Step 6   Modification of subroutine ENERGY:

- The only requirement is to modify the measure of integration similar to subroutine FLOW. Other terms remain unchanged (see Chapter 4 for derivation of the working equations of the scheme).

Step 7   Modification of subroutine STRESS:

- Find radial component of velocity (v_r) and (r) at the reduced integration points and calculate v_r/r. Include this term in the calculation of pressure via the penalty relation.

- Modify the measure of integration multiplying it to (r).

All of the described modifications are shown in the program listing. The rest of the subroutines will remain the same and no other modification is necessary. However, in practice a switch parameter can be defined in the program which by changing its value from 0 to 1 allows the user to select planar or axisymmetric options. Write formats used in the echo printing of input data and output files can also be modified to represent nodal coordinates, velocity and stress components in an (r, z) system instead of the planar forms given in the program listing.

7.5   CIRCULATORY FLOW IN A RECTANGULAR DOMAIN

As an example of the application of PPVN.f we consider simulation of the circulatory flow of low-density polyethylene melt in a rectangular domain of 0.05 m length and 0.01 m width. The flow regime is generated by the imposition of a steady motion at the top surface. The prescribed boundary conditions in this problem are as shown in Figure 7.2. Therefore at the side and bottom walls essential (Dirichlet) boundary conditions are given, no temperature condition at the top wall is equivalent to setting zero thermal stress (i.e. temperature gradients) at this boundary.

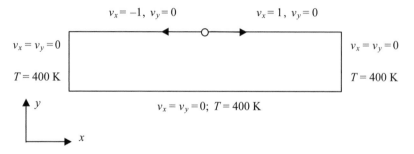

$v_x = -1, v_y = 0$ $v_x = 1, v_y = 0$

$v_x = v_y = 0$

$T = 400$ K

$v_x = v_y = 0$

$T = 400$ K

y

$v_x = v_y = 0; \ T = 400$ K

x

Figure 7.2   Boundary conditions in the sample simulation

Table 7.1 shows the structure of the input data file that is prepared according to the format described in the previous section.

Figures 7.3 and 7.4 show, respectively, the computed velocity and temperature fields, generated by PPVN.f for this example.

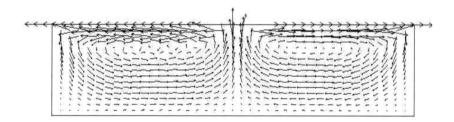

Figure 7.3   The simulated velocity field

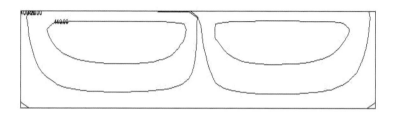

Figure 7.4   The simulated temperature contours

Table 7.1

Circulatory flow of low-density polyethylene melt in a rectangular domain

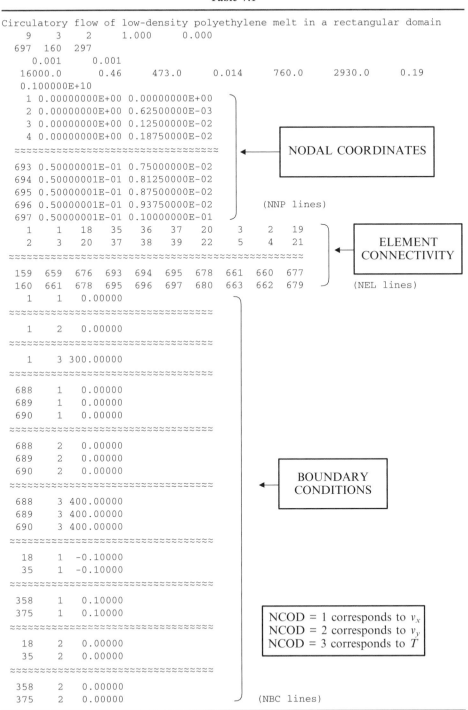

```
    9    3    2      1.000      0.000
  697  160  297
     0.001      0.001
  16000.0        0.46      473.0      0.014      760.0      2930.0      0.19
  0.100000E+10
    1 0.00000000E+00 0.00000000E+00 ⎫
    2 0.00000000E+00 0.62500000E-03 ⎪
    3 0.00000000E+00 0.12500000E-02 ⎪
    4 0.00000000E+00 0.18750000E-02 ⎪
≈≈≈≈≈≈≈≈≈≈≈≈≈≈≈≈≈≈≈≈≈≈≈≈≈≈≈≈≈≈≈≈≈ ⎬      NODAL COORDINATES
  693 0.50000001E-01 0.75000000E-02 ⎪
  694 0.50000001E-01 0.81250000E-02 ⎪
  695 0.50000001E-01 0.87500000E-02 ⎪
  696 0.50000001E-01 0.93750000E-02 ⎪     (NNP lines)
  697 0.50000001E-01 0.10000000E-01 ⎭
    1    1   18   35   36   37   20    3    2   19 ⎫
    2    3   20   37   38   39   22    5    4   21 ⎪
≈≈≈≈≈≈≈≈≈≈≈≈≈≈≈≈≈≈≈≈≈≈≈≈≈≈≈≈≈≈≈≈≈≈≈≈≈≈≈ ⎬      ELEMENT CONNECTIVITY
  159  659  676  693  694  695  678  661  660  677 ⎪
  160  661  678  695  696  697  680  663  662  679 ⎭  (NEL lines)
    1    1    0.00000 ⎫
≈≈≈≈≈≈≈≈≈≈≈≈≈≈≈≈≈≈≈≈≈≈≈≈≈≈≈≈≈≈≈≈≈≈ ⎪
    1    2    0.00000 ⎪
≈≈≈≈≈≈≈≈≈≈≈≈≈≈≈≈≈≈≈≈≈≈≈≈≈≈≈≈≈≈≈≈≈≈ ⎪
    1    3  300.00000 ⎪
≈≈≈≈≈≈≈≈≈≈≈≈≈≈≈≈≈≈≈≈≈≈≈≈≈≈≈≈≈≈≈≈≈≈ ⎪
  688    1    0.00000 ⎪
  689    1    0.00000 ⎪
  690    1    0.00000 ⎪
≈≈≈≈≈≈≈≈≈≈≈≈≈≈≈≈≈≈≈≈≈≈≈≈≈≈≈≈≈≈≈≈≈≈ ⎪
  688    2    0.00000 ⎪
  689    2    0.00000 ⎪
  690    2    0.00000 ⎬      BOUNDARY CONDITIONS
≈≈≈≈≈≈≈≈≈≈≈≈≈≈≈≈≈≈≈≈≈≈≈≈≈≈≈≈≈≈≈≈≈≈ ⎪
  688    3  400.00000 ⎪
  689    3  400.00000 ⎪
  690    3  400.00000 ⎪
≈≈≈≈≈≈≈≈≈≈≈≈≈≈≈≈≈≈≈≈≈≈≈≈≈≈≈≈≈≈≈≈≈≈ ⎪
   18    1   -0.10000 ⎪
   35    1   -0.10000 ⎪
≈≈≈≈≈≈≈≈≈≈≈≈≈≈≈≈≈≈≈≈≈≈≈≈≈≈≈≈≈≈≈≈≈≈ ⎪
  358    1    0.10000 ⎪
  375    1    0.10000 ⎪
≈≈≈≈≈≈≈≈≈≈≈≈≈≈≈≈≈≈≈≈≈≈≈≈≈≈≈≈≈≈≈≈≈≈ ⎪
   18    2    0.00000 ⎪
   35    2    0.00000 ⎪
≈≈≈≈≈≈≈≈≈≈≈≈≈≈≈≈≈≈≈≈≈≈≈≈≈≈≈≈≈≈≈≈≈≈ ⎪
  358    2    0.00000 ⎪
  375    2    0.00000 ⎭  (NBC lines)
```

NCOD = 1 corresponds to v_x
NCOD = 2 corresponds to v_y
NCOD = 3 corresponds to T

7.6 SOURCE CODE OF PPVN.F

This program has been originally developed and tested on a mainframe computer using the Unix operating system.

```
      PROGRAM PPVN
C
C ***********************************************************************
C
C A GALERKIN FINITE ELEMENT PROGRAM FOR THE SOLUTION OF STEADY-STATE
C NON-NEWTONIAN, NON-ISOTHERMAL PROBLEMS USING QUADRILATERAL
C       ISOPARAMETRIC ELEMENTS
C CONTINUOUS PENALTY SCHEME IN CONJUNCTION WITH SELECTIVELY REDUCED
C       INTEGRATION IS USED
C
C ***********************************************************************
C
C WORK FILES
C ==========
C   UNIT   CONTENTS
C ------------+----------------------------------------------------------
C      I
C   14  I  USED AS A WORK FILE IN THE SOLVER ROUTINE (SCRATCH FILE)
C      I
C   50  I  INPUT DATA CHANNEL
C      I
C   60  I  OUTPUT FILE FOR DOCUMENTATION
C      I
C   15  I  STORES SHAPE FUNCTIONS AND THEIR DERIVATIVES AT
C      I  'FULL' INTEGRATION POINTS(SCRATCH FILE)
C      I
C   16  I  STORES SHAPE FUNCTIONS AND THEIR DERIVATIVES AT
C      I  'REDUCED'INTEGRATION POINTS(SCRATCH FILE)
C      I
C ------------+----------------------------------------------------------
C
C MAIN VARIABLES
C =========
C   NODE (MAXEL, 18)  ELEMENT CONNECTIVITY ARRAY
C   CORD (MAXNP,NDIM)  NODAL COORDINATES ARRAY
C   AA  ( 18, 18)  ELEMENT COEFFICIENT MATRICES; FLOW EQUATIONS
C   RR  ( 18)     ELEMENT LOAD VECTOR; FLOW EQUATIONS
C   AE  ( 9, 9)  ELEMENT COEFFICIENT MATRICES; ENERGY EQUATION
C   RE  ( 9)     ELEMENT LOAD VECTOR; ENERGY EQUATION
C   VEL (MAXDF)     NODAL VELOCITIES
C   TEMP (MAXNP)    NODAL TEMPARATURES
C   STIFF(MAXAR)    GLOBAL STIFFNESS MATRIX
C   P  ( 9)     SHAPE FUNCTIONS
C   DEL ( 2, 9)  LOCAL DERIVATIVES OF SHAPE FUNCTIONS
C   B  ( 2, 9)  GLOBAL DERIVATIVES OF SHAPE FUNCTIONS
C   BC  (MAXDF)     BOUNDARY CONDITIONS ARRAY
C   VHEAT          GENERATED VISCOS HEAT
C   ALPHA          FACTOR FOR THE SELECTION OF SHEAR TERMS IN AA
C   BETA           FACTOR FOR THE SELECTION OF PENALTY TERMS IN AA
C   NNP            TOTAL NUMBER OF NODAL POINTS
C   NEL            TOTAL NUMBER OF ELEMENTS
```

```
C    NBC          TOTAL NUMBER OF BOUNDARY CONDITIONS
C    NDIM         DIMENSIONS OF THE SOLUTION DOMAIN
C    NDF          DEGREE OF FREEDOM PER NODE
C    TOLV         CONVERGENCE TOLERANCE PARAMETER FOR VELOCITIES
C    TOLT         CONVERGENCE TOLERANCE PARAMETER FOR TEMPERATURE
C    NUM          NUMBER OF INTEGRATION POINTS PER ELEMENT
C    RMAT1        MATERIAL PARAMETERS AT FULL INTEGRATION POINTS
C    RMAT2        MATERIAL PARAMETERS AT REDUCED INTEGRATION POINTS
C *******************************************************************
C
C *** STORAGE ALLOCATION
C
     PARAMETER(MAXEL=200,MAXNP=800,MAXBN=1200,MAXBC=300)
     PARAMETER(MAXDF=2000,MAXST=18)
C
C*** PARAMETERS SHOULD MATCH NUMBER OF ELEMENTS, NODES, ETC. USED IN A
PROBLEM
C
     IMPLICIT DOUBLE PRECISION (A-H,O-Z)
     DIMENSION TITLE ( 80)
     DIMENSION NODE (MAXEL, 18) ,PMAT (MAXEL,  8)
     DIMENSION CORD (MAXNP,  2)
     DIMENSION NCOD (MAXDF  ) ,BC  (MAXDF  )
     DIMENSION IBC  (MAXBC  ) ,JBC (MAXBC  ) ,VBC (MAXBC  )
     DIMENSION VEL  (MAXDF  ) ,R1  (MAXDF  ) ,TEMP (MAXNP  )
     DIMENSION CLUMP (MAXNP  ) ,STRES(MAXNP, 4)
     DIMENSION VET  (MAXDF  ) ,TET (MAXNP  )
     DIMENSION SINV (MAXEL, 13) ,NOPD (MAXEL, 18) ,RRSS (MAXDF  )
     DIMENSION AA  (  18, 18)
     DIMENSION AE  (  18, 18) ,RE  (    18)
     DIMENSION XG  (     3) ,CG  (     3)
     DIMENSION P   (    9) ,DEL ( 2,  9) ,B  ( 2,  9)
     DIMENSION RMAT1 (MAXEL, 13),RMAT2 (MAXEL, 13)
C
     CHARACTER*20 FILNAM
C
C *** GLOBAL STIFFNESS MATRIX
     COMMON/ONE/ STIFF(2000,300)
C
     PRINT*,'ENTER THE NAME OF YOUR DATA FILE'
     READ(*,111)FILNAM
 111 FORMAT(A40)
C
     OPEN(UNIT=60,FILE='SOL.OUT',FORM='FORMATTED',STATUS='NEW')
C
C *** INITIALIZE THE ARRAYS
C
     DO 9111 ITL = 1,MAXEL
     DO 9111 IVL = 1,18
              NODE (ITL,IVL) = 0
              NOPD (ITL,IVL) = 0
 9111 CONTINUE
     DO 9112 ITL = 1,MAXEL
     DO 9112 IVL = 1,8
              PMAT (ITL,IVL) = 0.0
 9112 CONTINUE
     DO 9113 ITL = 1,MAXNP
     DO 9113 IVL = 1,2
              CORD (ITL,IVL) = 0.0
```

OPEN & CLOSE
STATEMENTS
ARE USUALLY
SYSTEM
DEPENDENT

THE INITIALIZATION
STATEMENTS MAY
BE REDUNDANT IN
SOME SYSTEMS

```
 9113 CONTINUE
    DO 9114 ITL = 1,MAXNP
    DO 9114 IVL = 1,4
              STRES(ITL,IVL) = 0.0
 9114 CONTINUE
    DO 9115 ITL = 1,MAXEL
    DO 9115 IVL = 1,13
              SINV  (ITL,IVL)= 0.0
              RMAT1 (ITL,IVL)= 0.0
              RMAT2 (ITL,IVL)= 0.0
 9115 CONTINUE
    DO 9116 ITL = 1,MAXDF
              NCOD (ITL) = 0
              VEL  (ITL) = 0.0
              VET  (ITL) = 0.0
              R1   (ITL) = 0.0
              BC   (ITL) = 0.0
              RRSS (ITL) = 0.0
 9116 CONTINUE
    DO 9117 ITL = 1,MAXNP
              CLUMP (ITL) = 0.0
              TET   (ITL) = 0.0
              TEMP  (ITL) = 0.0
 9117 CONTINUE
    DO 9119 ITL = 1,MAXDF
    DO 9119 IVL = 1,MAXBN
              STIFF (IVL,ITL)= 0.0
 9119 CONTINUE
    DO 9121 ITL= 1,MAXBC
              IBC  (ITL) = 0
              JBC  (ITL) = 0
              VBC  (ITL) = 0.0
 9121 CONTINUE
C
C*** ARRAY SUBSCRIPTS AND THEIR ULTIMATE LIMITS
C
          NDIM = 2
          NUM  = 13
C
C ***********************************************************************
C
C  SET CONTROL PARAMETERS (DEFAULT VALUES ARE OVERWRITTEN BY INPUT DATA
C              IF SPECIFIED)
C   NCN      NUMBER OF NODES PER ELEMENT
C   NGAUS    NUMBER OF FULL INTEGRATION POINTS
C   MGAUS    NUMBER OF REDUCED INTEGRATION POINTS
C   NTER     MAXIMUM NUMBER OF INTEGRATIONS FOR NON-NEWTONIAN CASE
C
C ***********************************************************************
C
          NCN = 9
          NGAUS = 3
          MGAUS = 2
          NTER = 6
C
C *********** PARAMETERS FOR THE IDENTIFICATION OF PENALTY TERMS ******
C
          ALPHA = 1.0
          BETA = 0.0
```

```
C
   READ (50,1000) TITLE
   WRITE(60,2000) TITLE
C
C *** ELEMENT DESCRIPTION DATA
C
   READ (50,1010) NCNR,NGAUSR,MGAUSR,ALPHAR,BETAR
   IRED = ALPHAR
        IF(NCNR .NE.0 ) NCN  = NCNR
        IF(IRED .EQ.0 ) NUM  = 4
        IF(NCN  .NE.4 )  GO TO 4780
                  NGAUS = 2
                  MGAUS = 1
                  NUM  = 5
4780              CONTINUE
        IF(NGAUSR.NE.0 ) NGAUS = NGAUSR
        IF(MGAUSR.NE.0 ) MGAUS = MGAUSR
        IF(NGAUSR.NE.0 ) ALPHA = ALPHAR
        IF(MGAUSR.NE.0 ) BETA = BETAR
   WRITE(60,2010) NCN ,NGAUS ,MGAUS ,ALPHA ,BETA
C
C *** MESH DATA, BOUNDARY CONDITIONS AND TOLERANCE PARAMETERS
C
   READ (50,1020) NNP ,NEL ,NBC
C
        IF(NNP  .EQ.0 .OR.NNP .GT.MAXNP) GO TO 8000
        IF(NEL  .EQ.0 .OR.NEL .GT.MAXEL) GO TO 8000
        IF(NBC  .EQ.0 .OR.NBC .GT.MAXBC) GO TO 8000
   WRITE(60,2020) NNP ,NEL ,NBC
C
   READ(50,1030) TOLV ,TOLT
C
C ************************************************************************
C
 1000 FORMAT(80A1)
 1010 FORMAT(3I5,2F10.0)
 1020 FORMAT(3I5)
 1030 FORMAT(2F10.0)
 2000 FORMAT(' ',5(/),' ',20X,60('*'),/' ',20X,'*',58X,'*',/
    1' ',20X,'*',' A TWO-DIMENSIONAL FINITE ELEMENT MODEL OF A ',
    29X,'*',/' ',20X,'*',' NON-NEWTONIAN, NON-ISOTHERMAL FLOW USING ',
    3'REDUCED',9X,'*',/' ',20X,'*',' INTEGRATION / PENALTY FUNCTION ',
    4'METHOD.',18X,'*',/' ',20X,'*',58X,'*',/' ',20X,60('*')///,' ',
    520X,80('-'),/' ',20X,80A1,/' ',20X,80('-'),///)
 2010 FORMAT(' ',20X,3('['),' ELEMENT PRESCRIPTION ',10('.'),/
    125X,'NO.OF NODES PER ELEMENT          =',I10,/
    225X,'NO.OF INTEGRATION POINTS  (*FULL*)  =',I10,/
    325X,'NO.OF INTEGRATION POINTS (*REDUCED*)  =',I10,/
    425X,'SHEAR TERMS INTEGRATION FACTOR      =',F15.4,/
    525X,'PENALTY TERMS INTEGRATION FACTOR      =',F15.4,///)
 2020 FORMAT('0',20X,3('['),' MESH DATA PRESCRIPTION ',10('.'),/
    125X,'NO.OF NODAL POINTS            =',I10,/
    225X,'NO.OF ELEMENTS               =',I10,/
    325X,'NO.OF NODAL BOUNDARY CONDITIONS      =',I10,///)
C
C ************************************************************************
C
C   READ INPUT DATA FROM MAIN DATA FILE AND PREPARE ARRAYS FOR
C           SOLUTION PROCESS
```

```
C
C ***********************************************************************
C
    CALL GETMAT(NEL,PMAT,50,60,MAXEL,RTEM)
    CALL GETNOD(NNP,CORD,50,60,MAXNP,NDIM )
    CALL GETELM(NEL,NCN,NODE,50,60,MAXEL)
    CALL GETBCD(NBC,IBC,JBC,VBC,50,60,MAXBC)
C
C ***********************************************************************
C
C *INITIALIZE TEMPERATURE & SECOND INVARIANT OF RATE OF DEFORMATION TENSOR
C
                    DO 9996 IEL = 1,MAXEL
                    DO 9996 LG = 1, NUM
                    SINV(IEL,LG)= 0.250
 9996               CONTINUE
                    DO 9997 IVEL= 1,MAXDF
                    VEL (IVEL) =  0.0
 9997               CONTINUE
                    DO 9998 ITEM= 1,MAXNP
                    TEMP(ITEM) =  RTEM
 9998               CONTINUE
C
C*** MAIN SOLUTION LOOP
C
   DO 9999 ITER = 1 ,NTER
   PRINT*,'ITER=',ITER
C
   WRITE(60,2800) ITER
 2800 FORMAT(////' ',3('['),I5,'-TH ITERATION',10('.')//)
C
C * CALCULATE NODAL VELOCITIES ****************************************
C
                    REWIND 15
                    REWIND 16
              NDF  = 2
              NTOV = NDF * NNP
              NTRIX = NDF * NCN
   CALL CLEAN (R1   ,BC   ,NCOD ,NTOV ,MAXDF,MAXBN)
   CALL SETPRM (NNP ,NEL ,NCN ,NODE ,NDF ,MAXEL,MAXST)
   CALL PUTBCV (NNP ,NBC ,IBC ,JBC ,VBC ,NCOD ,MAXBC,MAXDF,BC)
C
   DO 5001 IEL=1,NEL
C
   CALL FLOW
   1 (NODE ,CORD ,PMAT ,NDF ,MAXBN,NCOD ,BC   ,VEL ,R1   ,RRSS
   2 ,TEMP ,NUM ,IEL ,ITER ,NEL ,NCN ,SINV ,NGAUS,MGAUS,P
   3 ,DEL ,B   ,ALPHA,BETA ,NTRIX,MAXEL,MAXNP,NOPD ,MAXST,MAXBC
   4 ,MAXDF,NDIM ,AA   ,XG   ,DA   ,NTOV ,IRED ,IBC ,JBC ,VBC
   5 ,NBC ,RMAT1,RMAT2)
C
 5001 CONTINUE
C
C ****** CHECK FOR CONVERGENCE ******************
C
   CALL CONTOL
   1(VEL,TEMP,ITER,NTOV,NNP,MAXNP,MAXDF,ERROV,ERROT,VET,TET)
C
         IF(ERROV.LT.TOLV.AND.ERROT.LT.TOLT) GO TO 88888
```

```
C
   GO TO 88880
C
88888 CALL OUTPUT (NNP ,VEL ,TEMP ,MAXDF,MAXNP)
C
   CALL CLEAN(R1,BC,NCOD,NTOV,MAXDF,MAXBN)
C
   GO TO 9000
C
88880 CONTINUE
C
C ************************************************************************
C
   CALL SECINV
   1  (NEL ,NNP ,NCN ,NGAUS,MGAUS,NODE ,SINV ,CORD ,P ,B ,
   2   DEL ,DA ,VEL ,MAXNP,MAXEL,MAXST,NDIM ,IRED ,NUM)
C
C *  CALCULATE NODAL TEMPERATURES ****************************************
C
                    REWIND 15
                    REWIND 16
              NDF = 1
              NTOV = NDF * NNP
              NTRIX= NDF * NCN
   CALL CLEAN (R1  ,BC  ,NCOD ,NTOV ,MAXDF,MAXBN)
   CALL SETPRM(NNP ,NEL ,NCN ,NODE ,NDF ,MAXEL,MAXST)
   CALL PUTBCT(NBC ,IBC ,JBC ,VBC ,NCOD ,BC  ,MAXBC,MAXDF)
C
   DO 5002 IEL=1,NEL
C
   CALL ENERGY
   1  (NODE ,CORD ,PMAT ,NDF ,MAXBN,NCOD ,BC  ,TEMP ,VEL ,RRSS
   2  ,R1 ,IRED ,XG ,NDIM ,DA ,IEL ,NEL ,NCN ,NTOV ,NUM
   3  ,ITER ,NGAUS,MGAUS,P ,DEL ,B ,SINV ,NTRIX,MAXEL,MAXNP
   4  ,MAXST,MAXDF,MAXBC,IBC ,JBC ,VBC ,NBC ,AE  ,RE  ,NOPD)
C
 5002 CONTINUE
C
 9999 CONTINUE
C
 9000 CONTINUE
C
C *** CALCULATION OF THE NODAL PRESSURE & STRESS USING VARIATIONAL RECOVERY
C
   CALL LUMPM
   1(CLUMP,NNP,MAXNP,NEL ,NGAUS,P ,DEL ,B ,MAXST,NODE,MAXEL,NCN)
C
   CALL STRESS
   1  (NEL ,NNP ,NCN ,NGAUS,MGAUS,NODE ,CORD ,P ,B ,DEL
   2  ,VEL ,MAXNP,MAXEL,MAXST,RMAT1,RMAT2,IRED ,STRES,CLUMP)
C
C
   CLOSE(UNIT=60)
   STOP
8000 CONTINUE
C
C ***********************************************************************
C
```

```
   WRITE(60,2995)
 2995 FORMAT('0',10('['),' INPUT DATA UNACCEPTABLE ',10('[')///)
   STOP
C
C ******************************************************************
C
   END
C
C ******************************************************************
C
   SUBROUTINE GAUSSP(NGAUS,XG,CG)
C
   IMPLICIT DOUBLE PRECISION(A-H,O-Z)
C
C *** NUMERICAL INTEGRATION POINT COORDINATES & WEIGHTS
C
   DIMENSION XG(3),CG(3)
   IF(NGAUS.NE.1) GO TO 10
   XG(1) = 0.0
   CG(1) = 2.0
   GO TO 300
 10 CONTINUE
   IF(NGAUS.NE.2) GO TO 100
   XG(1) = 0.57735026919D00
   XG(2) = -XG(1)
   CG(1) = 1.00
   CG(2) = 1.00
   GO TO 300
 100 CONTINUE
   XG(1) = 0.77459666924D00
   XG(2) = 0.0
   XG(3) = -XG(1)
   CG(1) = 0.55555555556D00
   CG(2) = 0.88888888889D00
   CG(3) = CG(1)
C
 300 RETURN
   END
C
C ******************************************************************
C
   SUBROUTINE SHAPE (G,H,P,DEL,NCN)
C
   IMPLICIT DOUBLE PRECISION(A-H,O-Z)
C
   DIMENSION P(9),DEL(2,9)
C
C  SHAPE FUNCTIONS & THEIR DERIVATIVES IN LOCAL COORDINATES ***
C***
C  1.BI-LINEAR -- FOUR-NODED QUADRILATERAL
C***
   GH = G*H
   IF(NCN.NE. 4) GO TO 9
   P(1) = (1.-G-H+GH)/4
   P(2) = (1.+G-H-GH)/4
   P(3) = (1.+G+H+GH)/4
   P(4) = (1.-G+H-GH)/4
   DEL(1,1) = -(1.-H)/4
   DEL(1,2) = (1.-H)/4
```

```
      DEL(1,3) = (1.+H)/4
      DEL(1,4) = -(1.+H)/4
      DEL(2,1) = -(1.-G)/4
      DEL(2,2) = -(1.+G)/4
      DEL(2,3) = (1.+G)/4
      DEL(2,4) = (1.-G)/4
      GO TO 30
    9 IF(NCN.NE.8) GO TO 10
      GG = G*G
      HH = H*H
      GGH = GG*H
      GHH = G*HH
C***
C 2.BI-QUADRATIC -- NINE-NODED QUADRILATERAL
C***
   10 G1=.5*G*(G-1.)
      G2=1.-G*G
      G3=.5*G*(G+1.)
      H1=.5*H*(H-1.)
      H2=1.-H*H
      H3=.5*H*(H+1.)
      P(1)=G1*H1
      P(2)=G2*H1
      P(3)=G3*H1
      P(4)=G3*H2
      P(5)=G3*H3
      P(6)=G2*H3
      P(7)=G1*H3
      P(8)=G1*H2
      P(9)=G2*H2
      DG1=G-.5
      DG2=-2.*G
      DG3=G+.5
      DH1=H-.5
      DH2=-2.*H
      DH3=H+.5
      DEL(1,1)=DG1*H1
      DEL(1,2)=DG2*H1
      DEL(1,3)=DG3*H1
      DEL(1,4)=DG3*H2
      DEL(1,5)=DG3*H3
      DEL(1,6)=DG2*H3
      DEL(1,7)=DG1*H3
      DEL(1,8)=DG1*H2
      DEL(1,9)=DG2*H2
      DEL(2,1)=G1*DH1
      DEL(2,2)=G2*DH1
      DEL(2,3)=G3*DH1
      DEL(2,4)=G3*DH2
      DEL(2,5)=G3*DH3
      DEL(2,6)=G2*DH3
      DEL(2,7)=G1*DH3
      DEL(2,8)=G1*DH2
      DEL(2,9)=G2*DH2
   30 CONTINUE
C
      RETURN
      END
C
```

```
C ***********************************************************************
C
   SUBROUTINE DERIV
   1 (IEL ,IG  ,JG  ,P  ,DEL ,B  ,NCN ,DA  ,CG  ,NOP
   2 ,CORD ,MAXEL,MAXNP,MAXST)
C
C*** JACOBIAN OF COORDINATES TRANSFORMATION & DERIVATIVES OF THE SHAPE
C   FUNCTIONS WRT GLOBAL VARIABLES
   IMPLICIT DOUBLE PRECISION(A-H,O-Z)
   DIMENSION P(9),B(2,9),DEL(2,9),CG(3),CJ(2,2),CJI(2,2)
   DIMENSION NOP(MAXEL,MAXST),CORD(MAXNP,2)
C
   DO 22 J=1,2
   DO 22 L=1,2
   GLSH=0.
   DO 21 K=1,NCN
   NN=IABS(NOP(IEL,K))
 21 GASH=GLSH+DEL(J,K)*CORD(NN,L)
 22 CJ(J,L)=GLSH
   DETJ=CJ(1,1)*CJ(2,2) -- CJ(1,2)*CJ(2,1)
   IF(DETJ) 29,33,29
 33 WRITE(60,34)
 34 FORMAT(1H ,22H DETJ=0 PROGRAM HALTED)
   STOP
C ***
 29 CJI(1,1) = CJ(2,2) / DETJ
   CJI(1,2) =-CJ(1,2) / DETJ
   CJI(2,1) =-CJ(2,1) / DETJ
   CJI(2,2) = CJ(1,1) / DETJ
C ***
   DO 40 J=1,2
   DO 40 L=1,NCN
   B(J,L)=0.0
   DO 40 K=1,2
 40 B(J,L) = B(J,L) + CJI(J,K) * DEL(K,L)
   DA = DETJ*CG(IG)*CG(JG)
C
   RETURN
   END
C
C ************************************************************************
C
   SUBROUTINE SECINV
   1 (NEL ,NNP ,NCN ,NGAUS,MGAUS,NODE ,SINV ,CORD ,P  ,B
   2 ,DEL ,DA  ,VEL ,MAXNP,MAXEL,MAXST,NDIM ,IRED ,NUM)
C
   IMPLICIT DOUBLE PRECISION(A-H,O-Z)
C
C FUNCTION
C --------
C   FINDS THE SECOND INVARIANT OF RATE OF DEFORMATION AT INTEGRATION
C   POINTS
C
   DIMENSION VEL (NNP , NDIM) ,CORD (MAXNP, NDIM)
   DIMENSION NODE (MAXEL,MAXST) ,SINV (MAXEL, NUM)
   DIMENSION P ( 9)    ,DEL ( 2, 9)
   DIMENSION B ( 2, 9)
C
C ***********************************************************************
```

```
C
                        REWIND 15
                        REWIND 16
   DO 5000 IEL = 1 , NEL
C
C ***
C
                IF(IRED.EQ.0) GO TO 5001
C
C *** FULL INTEGRATION POINTS ******************************************
C
        LG = 0
   DO 5010 IG = 1 ,NGAUS
   DO 5010 JG = 1 ,NGAUS
      LG = LG+1
                   READ (15) IIEL,IIG,JJG,P,DEL,B,DA
         X1 = 0.0
         U1 = 0.0
         U11 = 0.0
         U12 = 0.0
         U21 = 0.0
         U22 = 0.0
     DO 5020 ICN = 1 ,NCN
                   JCN = IABS(NODE(IEL,ICN))
```

> FOR *R,Z* OPTION FIND X1 & U1 AS
>
> X1 = X1 + P(ICN)*CORD(JCN,1)
> U1 = U1 + P(ICN)*CORD(JCN,1)
>
> IN THIS   DO LOOP

```
C
C *** COMPONENTS OF THE RATE OF DEFORMATION TENSOR
C
         U11 = U11 + B(1,ICN)* VEL(JCN,1)
         U12 = U12 + B(2,ICN)* VEL(JCN,1)
         U21 = U21 + B(1,ICN)* VEL(JCN,2)
         U22 = U22 + B(2,ICN)* VEL(JCN,2)
5020    CONTINUE
C
C *** SECOND INVARIANT OF THE RATE OF DEFORMATION TENSOR
C
   SINV(IEL,LG) = 0.125*((U11+U11)*(U11+U11)+
   1            (U12+U21)*(U12+U21)+
   2            (U21+U12)*(U21+U12)+
   3            (U22+U22)*(U22+U22))
C
5010 CONTINUE
C
5001 CONTINUE
C
C *** REDUCED INTEGRATION POINTS ******************************************
C
   IF(IRED .EQ.0) LG = 0
   DO 6010 IG = 1 ,MGAUS
   DO 6010 JG = 1 ,MGAUS
      LG = 1 + LG
                READ (16) IIEL,IIG,JJG,P,DEL,B,DA
```

> FOR *R,Z* OPTION MODIFY
> SINV(IEL,LG) = SINV(IEL,LG) +
>   0.5*(U1**2)/(X1**2)

```
              X1 = 0.0
              U1 = 0.0
              U11 = 0.0
              U12 = 0.0
              U21 = 0.0
              U22 = 0.0
          DO 6023 ICN = 1 ,NCN
                    JCN = IABS(NODE(IEL,ICN))
```

```
┌────────────────────────────────────────────────────────┐
│                                                          │
│   FOR R,Z OPTION FIND X1 & U1 AS                         │
│                                                          │
│       X1 = X1 + P(ICN)*CORD(JCN,1)                       │
│       U1 = U1 + P(ICN)*CORD(JCN,1)                       │
│                                                          │
│   IN THIS   DO LOOP                                      │
│                                                          │
└────────────────────────────────────────────────────────┘
```

```
C
C *** COMPONENTS OF THE RATE OF DEFORMATION TENSOR
C
          U11 = U11 + B(1,ICN)* VEL(JCN,1)
          U12 = U12 + B(2,ICN)* VEL(JCN,1)
          U21 = U21 + B(1,ICN)* VEL(JCN,2)
          U22 = U22 + B(2,ICN)* VEL(JCN,2)
6023 CONTINUE
C
C *** SECOND INVARIANT OF THE RATE OF DEFORMATION TENSOR
C
     SINV(IEL,LG) = 0.125*((U11+U11)*(U11+U11)+
    1              (U12+U21)*(U12+U21)+
    2              (U21+U12)*(U21+U12)+
    3              (U22+U22)*(U22+U22))
```

```
6010 CONTINUE
C
5000 CONTINUE
C
```

```
┌────────────────────────────────────────┐
│        FOR R,Z OPTION MODIFY             │
│   SINV(IEL,LG) = SINV(IEL,LG) +          │
│                0.5*(U1**2)/(X1**2)       │
└────────────────────────────────────────┘
```

```
     RETURN
     END
C
C ********************************************************************
C
     SUBROUTINE FLOW
    1 (NODE ,CORD ,PMAT ,NDF ,MAXBN,NCOD ,BC   ,VEL ,R1   ,RRSS
    2 ,TEMP ,NUM ,IEL ,ITER ,NEL ,NCN ,SINV ,NGAUS,MGAUS,P
    3 ,DEL ,B   ,ALPHA,BETA ,NTRIX,MAXEL,MAXNP,NOPD ,MAXST
    4 ,MAXBC,MAXDF,NDIM ,AA   ,XG   ,DA   ,NTOV ,IRED ,IBC ,JBC
    5 ,VBC ,NBC ,RMAT1,RMAT2)
C
C*** SOLUTION OF THE GENERALIZED NAVIER--STOKES EQUATION
C
     IMPLICIT DOUBLE PRECISION(A-H,O-Z)
C
     DIMENSION NODE (MAXEL,MAXST),PMAT (MAXEL,  8)
     DIMENSION CORD (MAXNP, NDIM)
     DIMENSION NCOD (MAXDF)    ,BC  (MAXDF)    ,SINV (MAXEL, NUM)
     DIMENSION VEL (MAXNP, NDIM),R1  (MAXDF)    ,TEMP (MAXNP)
     DIMENSION AA ( 18,   18),RR  ( 18)
     DIMENSION XG ( 3)    ,CG ( 3)
     DIMENSION X ( 2)    ,V ( 2)
```

```
      DIMENSION BICN (  2)    ,HH  (  2)
      DIMENSION P  (  9)  ,DEL (  2,  9),B  (  2,  9)
      DIMENSION IBC (MAXBC)   ,JBC (MAXBC)   ,VBC (MAXBC)
      DIMENSION RMAT1(MAXEL,  13),RMAT2(MAXEL,  13)
      DIMENSION NOPD (MAXEL,MAXST),RRSS (MAXDF)
      COMMON/ONE/ STIFF(2000,300)
C
C ***********************************************************************
C
                      RVISC = PMAT(IEL,1)
                      RPLAM = PMAT(IEL,2)
                      POWER = PMAT(IEL,3)
                      RTEM = PMAT(IEL,4)
                      TCO  = PMAT(IEL,5)
                      RODEN = PMAT(IEL,6)
         DO 5600 IDF = 1,NTRIX
         RR(IDF)   = 0.0
         DO 5600 JDF = 1,NTRIX
         AA(IDF,JDF) = 0.0
5600      CONTINUE
C
C ***
C
              IF(IRED.EQ.0) GO TO 5700
C
C *** 'FULL' INTEGRATION ********************************************
C
              CALL GAUSSP(NGAUS,XG,CG)
      LG=0
   DO 100 IG=1,NGAUS
              G = XG(IG)
   DO 100 JG=1,NGAUS
              H = XG(JG)
      LG=LG+1
                 IF(ITER.GT.1) GO TO 9996
   CALL SHAPE  (G,H,P,DEL,NCN)
   CALL DERIV
  1 (IEL,IG,JG,P,DEL,B,NCN,DA,CG,NODE,CORD,MAXEL,MAXNP,MAXST)
              WRITE(15) IEL ,IG ,JG ,P,DEL,B,DA
   GO TO 9995
 9996          READ (15) IIEL,IIG,JJG,P,DEL,B,DA
C
 9995 CONTINUE
C
C *** UPDATING VISCOSITY
C
   STEMP = 0.0
   DO  5333 IP = 1,NCN
       JP = IABS(NODE(IEL,IP))
           STEMP = STEMP + TEMP(JP) * P(IP)
 5333 CONTINUE
           EPSII = 1.D-10
           SRATE = SINV(IEL,LG)
           IF(SRATE.LT.EPSII) SRATE = EPSII
   CALL VISCA(RVISC,POWER,VISC,SRATE,STEMP,RTEM,TCO)
C
C *** CALCULATE VISCOSITY-DEPENDENT PENALTY PARAMETER
C
   PLAM = RPLAM * VISC
```

> FOR *R,Z* OPTION
> START A LOOP AND
> FIND X1 (SIMILAR
> TO SECINV) THEN
> MODIFY DA AS
> DA = DA*X(1)

```
C
   RMAT1(IEL,LG) = VISC
   RMAT2(IEL,LG) = PLAM
C
C *** ROW INDEX
C
   DO 19 I=1,NCN
              PICN  = P (I)
              BICN(1)= B(1,I)
              BICN(2)= B(2,I)
              PRVB  = RODEN *PICN
        J11 = I
        J12 = I + NCN
C
C *** COLUMN INDEX
C
   DO 19 J=1,NCN
        J21 = J
        J22 = J + NCN
C
C *** STIFFNESS MATRIX ***********************************************
C
   AA(J11,J21) = AA(J11,J21)
   1       + ALPHA*PRVB*(V(1)*B(1,J)+V(2)*B(2,J))*    DA
   2       + ALPHA*VISC*(2.0*B(1,I)*B(1,J)+B(2,I)*B(2,J))*DA
   3       + BETA *PLAM* B(1,I)*B(1,J) *DA
   AA(J11,J22) = AA(J11,J22)
   1       + ALPHA*VISC* B(2,I)*B(1,J) *DA
   2       + BETA *PLAM* B(1,I)*B(2,J) *DA
   AA(J12,J21) = AA(J12,J21)
   1       + ALPHA*VISC* B(1,I)*B(2,J) *DA
   2       + BETA *PLAM* B(2,I)*B(1,J) *DA
   AA(J12,J22) = AA(J12,J22)
   1       + ALPHA*PRVB*(V(1)*B(1,J)+V(2)*B(2,J))*    DA
   2       + ALPHA*VISC*(2.0*B(2,I)*B(2,J)+B(1,I)*B(1,J))*DA
   3       + BETA *PLAM* B(2,I)*B(2,J) *DA

19  CONTINUE
C
100  CONTINUE
C
5700 CONTINUE
C
C *** 'REDUCED' INTEGRATION **
C
   IF(IRED .EQ.0) LG = 0
                  CALL GAUSSP(MGAUS,XG,CG)
   DO 200 IG=1,MGAUS
                  G = XG(IG)
   DO 200 JG=1,MGAUS
                  H = XG(JG)
   LG=LG+1
                  IF(ITER.GT.1) GO TO 9994
   CALL SHAPE (G,H,P,DEL,NCN)
   CALL DERIV
   1 (IEL,IG,JG,P,DEL,B,NCN,DA,CG,NODE,CORD,MAXEL,MAXNP,MAXST)
                  WRITE(16) IEL ,IG ,JG ,P,DEL,B,DA
   GO TO 9993
9994              READ (16) IIEL,IIG,JJG,P,DEL,B,DA
```

Box (upper right):

> THIS TERM SHOULD BE MULTIPLIED BY ZERO FOR STOKES FLOW CALCULATIONS

Box (lower right):

> FOR R,Z OPTION MODIFY AS
>
> AA(J11,J21) = AA(J11,J21) + BETA *PLAM*(B(1,I)*P(J)/X(1) + B(1,J)*P(I)/X(1) + P(I)*P(J)/X(1)**2)DA
>
> AA(J11,J22) = AA(J11,J22) + BETA* PLAM*P(I)*B(2,J)/X(1)*DA
>
> AA(J12,J21) = AA(J12,J21) + BETA* PLAM*P(J)*B(2,I)/X(1)*DA

```
C
 9993 CONTINUE
C
C *** UPDATING VISCOSITY
C
   STEMP = 0.0
   DO 3334 IP = 1,NCN
       JP = IABS(NODE(IEL,IP))
           STEMP = STEMP + TEMP(JP) * P(IP)
 3334 CONTINUE
           EPSII = 1.D-10
           SRATE = SINV(IEL,LG)
           IF(SRATE.LT.EPSII) SRATE = EPSII
   CALL VISCA(RVISC,POWER,VISC,SRATE,STEMP,RTEM,TCO)
C
C *** CALCULATE VISCOSITY-DEPENDENT PENALTY PARAMETER
C
   PLAM = RPLAM * VISC
C
   RMAT1(IEL,LG) = VISC
   RMAT2(IEL,LG) = PLAM
C
C *** ROW INDEX
C
   DO 20 I=1,NCN
           PICN = P (I)
           BICN(1)= B(1,I)
           BICN(2)= B(2,I)

           PRVB = RODEN * PICN
       J11 = I
       J12 = I + NCN
C
C *** COLUMN INDEX
C
   DO 20 J=1,NCN
           J21 = J
           J22 = J+NCN
C
C *** STIFFNESS MATRIX FOR 'REDUCED' INTEGRATION **********************
C
   AA(J11,J21) = AA(J11,J21)
   1       + (1.0-ALPHA)*PRVB*(V(1)*B(1,J)+V(2)*B(2,J))*    DA
   2       + (1.0-ALPHA)*VISC*(2.0*B(1,I)*B(1,J)+B(2,I)*B(2,J))*DA
   3       + (1.0-BETA) *PLAM* B(1,I)*B(1,J) *DA
   AA(J11,J22) = AA(J11,J22)
   1       + (1.0-ALPHA)*VISC* B(2,I)*B(1,J) *DA
   2       + (1.0-BETA) *PLAM* B(1,I)*B(2,J) *DA
   AA(J12,J21) = AA(J12,J21)
   1       + (1.0-ALPHA)*VISC* B(1,I)*B(2,J) *DA
   2       + (1.0-BETA) *PLAM* B(2,I)*B(1,J) *DA
   AA(J12,J22)=AA(J12,J22)
   1       + (1.0-ALPHA)*PRVB*(V(1)*B(1,J)+V(2)*B(2,J))*    DA
   2       + (1.0-ALPHA)*VISC*(2.0*B(2,I)*B(2,J)+B(1,I)*B(1,J))*DA
   3       + (1.0-BETA) *PLAM* B(2,I)*B(2,J) *DA
 20   CONTINUE
 200   CONTINUE
```

> FOR *R,Z* OPTION START A LOOP AND FIND X1 (SIMILAR TO SECINV) THEN MODIFY DA AS DA = DA*X(1)

> THIS TERM SHOULD BE MULTIPLIED BY ZERO FOR STOKES FLOW CALCULATIONS

> FOR *R,Z* OPTION REPEAT SIMILAR MODIFICATIONS AS THE FULL INTEGRATION USING (1.0-BETA)

```
      CALL SOLVER
     1(AA   ,RR   ,IEL ,NODE ,NCN  ,IBC  ,JBC  ,VBC  ,NBC  ,BC   ,NTOV
     2,NCOD ,NTRIX,NEL ,VEL  ,R1   ,MAXEL,MAXDF,MAXST,MAXBC,MAXBN,NDF
     3,NOPD ,RRSS)
C
      RETURN
      END
C
C **************************************************************************
C
      SUBROUTINE ENERGY
     1 (NODE ,CORD ,PMAT ,NDF ,MAXBN,NCOD ,BC   ,TEMP ,VEL ,RRSS
     2 ,R1   ,IRED ,XG   ,NDIM ,DA   ,IEL ,NEL ,NCN  ,NTOV ,NUM
     3 ,ITER ,NGAUS,MGAUS,P    ,DEL ,B   ,SINV ,NTRIX,MAXEL,MAXNP
     4 ,MAXST,MAXDF,MAXBC,IBC  ,JBC ,VBC ,NBC  ,AE   ,RE   ,NOPD)
C
C*** SOLUTION OF THE ENERGY EQUATION
C
      IMPLICIT DOUBLE PRECISION(A-H,O-Z)
C
      DIMENSION NODE (MAXEL,MAXST),CORD (MAXNP, NDIM)
      DIMENSION NCOD (MAXDF)      ,BC   (MAXDF)     ,SINV (MAXEL, NUM)
      DIMENSION TEMP (MAXNP)      ,R1   (MAXDF)     ,VEL (MAXNP, NDIM)
      DIMENSION AE ( 18,  18),RE ( 18)
      DIMENSION XG ( 3)
      DIMENSION P ( 9)   ,DEL ( 2, 9),B ( 2, 9)
      DIMENSION X ( 2)   ,V ( 2)
      DIMENSION BICN ( 2)   ,BJCN ( 2)
      DIMENSION HH ( 2) ,HD(2)  ,PMAT (MAXEL, 8)
      DIMENSION IBC (MAXBC)   ,JBC (MAXBC)   ,VBC (MAXBC)
      DIMENSION NOPD (MAXEL,MAXST),RRSS (MAXDF)
C
      COMMON/ONE/ STIFF(2000,300)
C
C **************************************************************************
C
                       RVISC = PMAT(IEL,1)
                       RPLAM = PMAT(IEL,2)
                       POWER = PMAT(IEL,3)
                       RTEM  = PMAT(IEL,4)
                       TCO   = PMAT(IEL,5)
                       RODEN = PMAT(IEL,6)
C
C *** BASIC ELEMENT LOOP ***********************************************
C
              DO 4900 ITRIX = 1 ,NTRIX
              RE(ITRIX)   = 0.0
              DO 4900 JTRIX = 1 ,NTRIX
              AE(ITRIX,JTRIX) = 0.0
 4900           CONTINUE
C
C *** NUMERICAL INTEGRATION
C
      MO = NGAUS
      IF(IRED .EQ.0 ) MO = MGAUS
           LG = 0
        DO 5010 IG = 1 ,MO
        DO 5010 JG = 1 ,MO
           LG = LG +1
```

```
C
C SHAPE FUNCTIONS & THEIR CARTESIAN DERIVATIVES ARE READ FROM A WORK FILE
C
   IF(IRED .EQ.1) READ (15) IIEL,IIG,JJG,P,DEL,B,DA
   IF(IRED .EQ.0) READ (16) IIEL,IIG,JJG,P,DEL,B,DA
C
C *** UPDATING VISCOSITY
C
   STEMP = 0.0
   DO  3337 IP = 1,NCN
        JP = IABS(NODE(IEL,IP))
             STEMP = STEMP + TEMP(JP) * P(IP)
 3337 CONTINUE
             EPSII = 1.D-10
             SRATE = SINV(IEL,LG)
             IF(SRATE.LT.EPSII) SRATE = EPSII
C
   CALL VISCA(RVISC,POWER,VISC,SRATE,STEMP,RTEM,TCO)
   CP  =PMAT(IEL,7)
   CONDK=PMAT(IEL,8)
C
C *** CALCULATE VISCOUS HEAT DISSIPATION
C
   VHEAT = 4. * VISC * SRATE
C
C *** COEFFICIENTS EVALUATED AT THE INTEGRATION POINTS
C
                DO 5030 IDF = 1 , 2
                X(IDF)   = 0.0
                V(IDF)   = 0.0
                HD(IDF)  = 0.0
 5030           CONTINUE
     DO 5040 ICN = 1 ,NCN
        JCN = IABS(NODE(IEL,ICN))
     DO 5040 IDF = 1 , 2
     X(IDF)   = X(IDF) + P(ICN)*CORD(JCN,IDF)
     V(IDF)   = V(IDF) + P(ICN)*VEL (JCN,IDF)
     DO 5040 JDF = 1 ,2
     HD(IDF)   = HD(IDF)+ 2.0*DEL(JDF,ICN)*CORD(JCN,IDF)
 5040   CONTINUE
C
C *** STREAMLINE UPWINDING - CALCULATION OF UPWINDING PARAMETER
C
        HDD     = SQRT(HD(1)**2 + HD(2)**2)
        AVV     = SQRT( V(1)**2 + V(2)**2)
C ***
        IF(AVV .LT. 1.D-10)
   1    AVV     = 1.D-10
        CONST   = 0.5 * HDD / AVV
C
   DO 6000 ICN = 1 ,NCN
                PICN = P (ICN)
                BICN(1)= B(1,ICN)
                BICN(2)= B(2,ICN)
C *** CALCULATE UPWINDED WEIGHT FUNCTION
        DO 4850 JDF = 1 , 2
        PICN    = PICN + V(JDF)*CONST*BICN(JDF)
 4850     CONTINUE
C
```

> FOR *R,Z* OPTION FIND (X1) AT THE INTEGRATION POINTS AND MODIFY DA (SIMILAR TO SUBROUTINE **FLOW**)

```
C *** ROW INDEX
C
                    IR = ICN
C
C *** SOURCE FUNCTION
C
   RE(IR)    = RE(IR)    + P(ICN)*VHEAT*DA
C
   DO 6010 JCN = 1 ,NCN
                  PJCN  = P (JCN)
                  BJCN(1)= B(1,JCN)
                  BJCN(2)= B(2,JCN)
C
C *** COLUMN INDEX
C
                    IC = JCN
C
               DO 6020 MDF = 1 ,2
C
C *** DIAGONAL ENTRY *** CONVECTION AND DIFFUSION TERMS
C
   AE (IR,IC) = AE(IR,IC) + RODEN*CP*PICN*V(MDF)*BJCN(MDF)*DA
   1              +    CONDK*BICN(MDF)*BJCN(MDF)*DA
C
6020              CONTINUE
C
6010 CONTINUE
6000 CONTINUE
C
 5010 CONTINUE
C
C *** ASSEMBLE AND SOLVE
C
   CALL SOLVER
   1(AE  ,RE  ,IEL ,NODE ,NCN ,IBC ,JBC ,VBC ,NBC ,BC  ,NTOV
   2,NCOD ,NTRIX,NEL ,TEMP ,R1  ,MAXEL,MAXDF,MAXST,MAXBC,MAXBN,NDF
   3,NOPD ,RRSS)
C
C *** END OF BASIC ELEMENT LOOP *****************************************
C
C
   RETURN
   END
C
C **********************************************************************
C
   SUBROUTINE STRESS
   1  (NEL ,NNP ,NCN ,NGAUS,MGAUS,NODE ,CORD ,P  ,B  ,DEL
   2  ,VEL ,MAXNP,MAXEL,MAXST,RMAT1,RMAT2,IRED ,STRES,CLUMP)
C
   IMPLICIT DOUBLE PRECISION(A-H,O-Z)
C
C FUNCTION
C --------
C   CALCULATES PRESSURE AND STRESS COMPONENTS AT 'REDUCED', INTEGRATION
C   POINTS AND WRITES INTO OUTPUT FILE.
C   VARIATIONAL RECOVERY OF PRESSURE, AND STRESS COMPONENTS AT NODES
C
```

```
   DIMENSION NODE (MAXEL,MAXST) ,CORD (MAXNP,  2)
   DIMENSION RMAT1(MAXEL,  13) ,RMAT2(MAXEL,  13)
   DIMENSION P  (  9)    ,DEL (  2,  9)
   DIMENSION B  (  2,  9)
   DIMENSION STRES (NNP ,   4) ,CLUMP(MAXNP)
C
C  *********************************************************************
C
                        REWIND 16
                     DO 4990 INP = 1,NNP
DO 4990 ICP  = 1, 4
STRES(INP,ICP) = 0.0
 4990                    CONTINUE
   DO 5000 IEL = 1 ,NEL
      NG = 0
   DO 6010 IG = 1 ,MGAUS
   DO 6010 JG = 1 ,MGAUS
      NG = 1 + NG
                 READ (16) JEL,KG,LG,P,DEL,B,DA
   IFG = NG +IRED*(NGAUS**2)
   RVISC=RMAT1(IEL,IFG)
   RPLAM=RMAT2(IEL,IFG)
      XG1 = 0.0
      XG2 = 0.0
      U11 = 0.0
      U12 = 0.0
      U21 = 0.0
      U22 = 0.0
   DO 6020 ICN = 1 ,NCN
            JCN = IABS(NODE(IEL,ICN))
      XG1 = XG1 + P(ICN)*CORD(JCN,1)
      XG2 = XG2 + P(ICN)*CORD(JCN,2)
      U11 = U11 + B(1,ICN)*VEL(JCN,1)
      U12 = U12 + B(2,ICN)*VEL(JCN,1)
      U21 = U21 + B(1,ICN)*VEL(JCN,2)
      U22 = U22 + B(2,ICN)*VEL(JCN,2)
6020    CONTINUE
C
C *** CARTESIAN COMPONENTS OF THE STRESS TENSOR
C
   PRES =-RPLAM * (U11 + U22)
   SD11 = 2.0  *RVISC * U11
   SD12 = RVISC * (U12 + U21)
   SD22 = 2.0  *RVISC * U22
C
   S11 =-PRES + SD11
   S12 = SD12
   S22 =-PRES + SD22
C
C * CALCULATE PRESSURE & STRESS AT NODAL POINTS *(VARIATIONAL RECOVERY)*
C
      DO 6500 ICN = 1 ,NCN
         JCN = IABS(NODE(IEL,ICN))
      STRES(JCN,1)= STRES(JCN,1)
   1        + P(ICN)*PRES *DA / CLUMP(JCN)
      STRES(JCN,2)= STRES(JCN,2)
   1        + P(ICN)*S11 *DA / CLUMP(JCN)
```

> DA IS READ FROM A FILE; IT IS ALREADY MODIFIED IF YOU ARE USING *R,Z* OPTION

> FOR *R,Z* OPTION FIND U1 IN THIS LOOP AS U1 = U1 + P(ICN)*VEL(JCN,1)

> FOR *R,Z* OPTION FIND VR = U1/XG1 AND MODIFY PRES AS PRES = PRES – RPLAM&VR

```
              STRES(JCN,3)= STRES(JCN,3)
     1                + P(ICN)*S22 *DA / CLUMP(JCN)
              STRES(JCN,4)= STRES(JCN,4)
     1                + P(ICN)*S12 *DA / CLUMP(JCN)
 6500         CONTINUE
 6010 CONTINUE
C
 5000 CONTINUE
C **********************************************************************
   WRITE(60,2100)
 2100 FORMAT('1',' ***VARIATIONAL RECOVERY***',/
   1/' NODE',11X,'PRES',12X,'S11',12X,'S22',12X,'S12')
   WRITE(60,2110) (INP,(STRES(INP,ICP),ICP = 1,4),INP = 1,NNP)
 2110 FORMAT(I5,4E15.4)
C
   RETURN
   END
C
C **********************************************************************
C
   SUBROUTINE LUMPM
   1(CLUMP,NNP ,MAXNP,NEL ,NGAUS,P ,DEL ,B ,MAXST,NODE ,MAXEL,NCN)
C
   IMPLICIT DOUBLE PRECISION(A-H,O-Z)
C
   DIMENSION B  (  2,  9) ,DEL (  2,  9) ,P  (  9)
   DIMENSION CLUMP(MAXNP)
   DIMENSION NODE (MAXEL,MAXST)
                  DO 5000 INP = 1 ,NNP
                  CLUMP (INP)= 0.0
 5000              CONTINUE
        REWIND 15
C **********************************************************************
        DO 5010       IEL = 1 ,NEL
        DO 5020       IG = 1 ,NGAUS
        DO 5020       JG = 1 ,NGAUS
        READ (15) JEL ,KG ,LG ,P ,DEL ,B ,DA
           DO 5030  ICN = 1 ,NCN
                WW = 0.0
           DO 5040  JCN = 1 ,NCN
                WW = WW + P(ICN)*P(JCN)*DA
 5040         CONTINUE
           INP       = IABS(NODE(IEL,ICN))
           CLUMP(INP)    =CLUMP(INP) + WW
 5030         CONTINUE
 5020         CONTINUE
 5010         CONTINUE
   RETURN
   END
C
C **********************************************************************
C
   SUBROUTINE SOLVER
   1(RELST,RELRH,MM   ,NOP ,NCN ,IBC ,JBC ,VBC ,NB C ,RBC ,NTOV
   2,NCOD ,NTRIX,NE   ,RSOLN,RRHS ,MAXEL,MAXDF,MAXST,MAXBC,MAXBN,NDF
   3,NOPD ,RRSS)
C
C **********************************************************************
C
```

> STORE THE
> RESULTS FOR
> POST-PROCESSING

```
      IMPLICIT DOUBLE PRECISION(A-H,O-Z)
C
C*** THIS SUBROUTINE ASSEMBLES AND SOLVES GLOBAL STIFFNESS EQUATIONS
C
C **********************************************************************
C   ARGUMENTS
C   ---------
C
C   RELST (MAXST,MAXST) ELEMENT COEFFICIENT MATRICES
C   RELRH (MAXST)       ELEMENT LOAD VECTOR
C   NOP   (MAXEL,MAXST) ELEMENT CONNECTIVITY
C   RSOLN (MAXDF)       NODAL VELOCITIES
C   RRHS  (MAXDF)       GLOBAL LOAD VECTOR
C   STIFF (MAXAR)       GLOBAL STIFFNESS MATRIX
C   RBC   (MAXDF)       ARRAY FOR SORTING BOUNDARY CONDITIONS
C   NE            TOTAL NUMBER OF ELEMENTS IN THE MESH
C   IBC   (MAXBC)       ARRAY FOR BOUNDARY NODES
C   JBC   (MAXBC)       ARRAY FOR DEGREES OF FREEDOM CORRESPONDING TO
C                 A BOUNDARY CONDITION
C   VBC   (MAXBC)       ARRAY FOR BOUNDARY CONDITION VALUES
C
C **********************************************************************
C
      DIMENSION RELST (MAXST,MAXST),RELRH (MAXST)
      DIMENSION NOP   (MAXEL,MAXST)
      DIMENSION RRHS  (MAXDF)
      DIMENSION RSOLN (MAXDF)
      DIMENSION NCOD  (MAXDF)
      DIMENSION RBC   (MAXDF)
      DIMENSION IBC   (MAXBC)
      DIMENSION JBC   (MAXBC)
      DIMENSION VBC   (MAXBC)
      DIMENSION NOPD (MAXEL,MAXST),RRSS (MAXDF)
C
      COMMON/ONE/ STIFF(2000,300)
C
C **********************************************************************
C
      IF(MM .EQ. 1) REWIND 14
              WRITE(14) RELRH
C
C
      IF(MM .LT. NE) RETURN
C
      DO 5000 I=1,NE
      DO 5000 J=1,NTRIX
      NOPD(I,J)=NOP(I,J)
 5000     CONTINUE
C
                  REWIND 14
C
      CALL MODIFY
     1  (NOP ,NE   ,NCN ,IBC ,JBC ,RBC,NBC,VBC,NCOD,NTOV,
     2   MAXEL,MAXDF,MAXST,MAXBC,NDF)
C
      CALL BANDWD
     1  (NOP ,NE   ,IBAND,NTRIX,MAXST,MAXEL)
C
              DO 5010 IMM=1,NE
```

```
C
                    READ(14) RELRH
C
   CALL ASSEMB
   1  (IBAND,NTRIX,IMM,NOP,NTOV,RELRH,RELST,RRHS,MAXEL,MAXST)
C
 5010               CONTINUE
C
   CALL SOLVE
   1  (RRHS,MAXBN,RSOLN,IBAND,NTOV,NCOD,RBC,MAXDF,NDF,RRSS)
C
   DO 5020 I=1,NE
   DO 5020 J=1,NTRIX
    NOP(I,J)=NOPD(I,J)
 5020      CONTINUE
C
   RETURN
   END
C
C ***********************************************************************
C
   SUBROUTINE ASSEMB
   1  (IBAND,NTRIX,MM,NOP,NTOV,RELRH,RELST,RRHS ,MAXEL,MAXST)
C
   IMPLICIT DOUBLE PRECISION(A-H,O-Z)
C
C FUNCTION
C --------
C   ASSEMBLES THE ELEMENTAL STIFFNESS MATRICES
C
   DIMENSION RELST (MAXST,MAXST),RELRH (MAXST)
   DIMENSION RRHS (NTOV )
   DIMENSION NOP  (MAXEL,MAXST)
C
   COMMON/ONE/ STIFF(2000,300)
C
C ***********************************************************************
C
C*** CALCULATE HALF BANDWIDTH PARAMETERS
C
   IHBW1=(IBAND+1)/2
C
C*** LOOP THROUGH ROWS OF ELEMENT STIFFNESS MATRICES
C
   DO 5000 ITRIX=  1  ,NTRIX
       IROW =NOP(MM  ,ITRIX)
C
C*** ASSEMBLE RIGHT-HAND SIDE
C
   RRHS(IROW) = RRHS(IROW) + RELRH(ITRIX)
C
C*** LOOP THROUGH COLUMNS OF ELEMENT STIFFNESS MATRICES
C
   DO 5000 JTRIX=  1  ,NTRIX
       JCOLM=NOP(MM  ,JTRIX)
C
C*** ASSEMBLE GLOBAL STIFFNESS MATRIX IN A BANDED FORM
C
   JBAND=JCOLM-IROW+IHBW1
```

```
      STIFF(IROW,JBAND) = STIFF(IROW,JBAND) + RELST(ITRIX,JTRIX)
 5000 CONTINUE
C
      RETURN
      END
C
C ************************************************************************
C
      SUBROUTINE BANDWD
     1  (NOP ,NE   ,IBAND,NTRIX,MAXST,MAXEL)
C
      IMPLICIT DOUBLE PRECISION(A-H,O-Z)
C
C FUNCTION
C --------
C    FINDS THE MAXIMUM BANDWIDTH IN THE ASSEMBLED GLOBAL MATRIX
C
      DIMENSION NOP (MAXEL,MAXST)
C
C ************************************************************************
C
      DO 5000  MM=1,NE
         NPMAX=NOP(MM,1)
         NPMIN=NOP(MM,1)
      DO 5001 JTRIX=2,NTRIX
      IF(NOP(MM,JTRIX).GT.NPMAX) NPMAX=NOP(MM,JTRIX)
      IF(NOP(MM,JTRIX).LT.NPMIN) NPMIN=NOP(MM,JTRIX)
 5001 CONTINUE
      NDELT=NPMAX-NPMIN
      MBAND=2*NDELT+1
      IF(MM   .EQ.   1) IBAND=MBAND
      IF(IBAND .LT. MBAND) IBAND=MBAND
 5000 CONTINUE
      WRITE(60,2000)IBAND
 2000 FORMAT(1H ,7H IBAND=,I5)
 C
      RETURN
      END
C
C ************************************************************************
C
      SUBROUTINE SOLVE
     1   (RRHS,MAXBN,RSOLN,IBAND,NTOV,NCOD,RBC,MAXDF,NDF,RRSS)
C
      IMPLICIT DOUBLE PRECISION(A-H,O-Z)
C
C FUNCTION
C --------
C    SOLVES THE GLOBAL STIFFNESS MATRIX USING LU DECOMPOSITION
C
      DIMENSION RRHS (NTOV )
      DIMENSION RSOLN (NTOV )
      DIMENSION NCOD (MAXDF)
      DIMENSION RBC  (MAXDF)
      DIMENSION RRSS (MAXDF)
C
      COMMON/ONE/ STIFF(2000,300)
C
C ************************************************************************
```

```
C
   DO 1021 IMBO = 1,MAXDF
   RRSS  (IMBO)= 0.0
 1021 CONTINUE
C
C*** CALCULATE HALF BANDWIDTH PARAMETERS
C
   IHBW=(IBAND-1)/2
   IHBW1=IHBW+1
C
C*** BOUNDARY CONDITIONS
C
   DO 4995 ITOV=1,NTOV
   IF(NCOD(ITOV).NE.1) GO TO 4994
C
C*** INSERT BOUNDARY CONDITIONS
C
   STIFF(ITOV,IHBW1)=STIFF(ITOV,IHBW1)*1.D10
C
C*** MODIFY RHS VECTOR
C
   RRHS(ITOV)=RRHS(ITOV)+STIFF(ITOV,IHBW1)*RBC(ITOV)
   go to 4995
4994 if(ncod(itov).eq.0) go to 4995
4995 CONTINUE
C
C***   LU DECOMPOSITION
C
C*** SET UP THE FIRST ROW.
C
   DO 5000 LFIRST=1,IHBW
   MFIRST=LFIRST+1
   NFIRST=IHBW1-LFIRST
   STIFF(MFIRST,NFIRST)=STIFF(MFIRST,NFIRST)/STIFF(1,IHBW1)
 5000 CONTINUE
C
C*** COMPLETE LU DECOMPOSITION FOR INTERIOR ELEMENTS
C
   DO   5001   ITOV = 2 ,NTOV
           JTOV =-1
           KTOV = ITOV+IHBW -1
   IF(KTOV.GT.NTOV) KTOV = NTOV
           LTOV = ITOV-IHBW1
   DO   5002   MTOV = ITOV,KTOV
           JTOV = JTOV+1
           K1  = ITOV-IHBW +JTOV
   IF(K1 .LT. 1)  K1  = 1
           LCOLM= MTOV-LTOV
           II1 = ITOV-1
   DO   5002   KROW = K1 ,II1
           K2  = KROW-IHBW1
           MCOLM= KROW-LTOV
           NCOLM= MTOV-K2
   STIFF(ITOV,LCOLM) =
   1  STIFF(ITOV,LCOLM)-STIFF(ITOV,MCOLM)*STIFF(KROW,NCOLM)
 5002                         CONTINUE
   IF(ITOV.GE.NTOV) GO TO 5001
           JTOV = 0
           KTOV = KTOV+1
```

```fortran
      IF(KTOV.GT.NTOV) KTOV = NTOV
            II2 = ITOV+1
   DO   5003   IROW = II2 ,KTOV
            JTOV = JTOV+1
            K1   = ITOV-IHBW +JTOV
   IF(K1 .LT.  1) K1  = 1
            J2   = IROW-IHBW1
            L  = ITOV-J2
   IF(JTOV.GE.IHBW) GO TO 8000
   DO   5004   K  = K1 ,II1
            K2   = K  -IHBW1
            M  = K  -J2
            N  = ITOV-K2
   STIFF(IROW,L)=STIFF(IROW,L)-STIFF(IROW,M)*STIFF(K,N)
 5004                       CONTINUE
 8000                       CONTINUE
   STIFF(IROW,L)=STIFF(IROW,L)/STIFF(ITOV,IHBW1)
 5003                       CONTINUE
 5001                       CONTINUE
C
C***  FORWARD REDUCTION LY=F
C
   DO   5005   I  = 2 ,NTOV
            I1  = I  -IHBW
   IF(I1 .LT.  1) I1  = 1
            I2  = I  -IHBW1
            II1 = I  -1
   DO   5006   K  = I1 ,II1
            L  = K  -I2
   RRHS(I)=RRHS(I)-STIFF(I,L)*RRHS(K)
 5006                       CONTINUE
 5005                       CONTINUE
C
C*** FIND THE SOLUTION VECTOR BY BACK SUBSTITUTION UX=Y
C
   RSOLN(NTOV)=RRHS(NTOV)/STIFF(NTOV,IHBW1)
   DO   5007   I  = 2 ,NTOV
            II  = NTOV-I +1
            I1  = II +IHBW
   IF(I1 .GT.NTOV) I1  = NTOV
            I2   = II -IHBW1
            II1 = II +1
   DO   5008   K  = II1 ,I1
            L  = K  -I2
   RRHS (II)=RRHS(II)-STIFF(II,L)*RSOLN(K)
 5008                       CONTINUE
   RSOLN(II)=RRHS(II)/STIFF(II,IHBW1)
 5007                       CONTINUE
   IF(NDF .EQ. 1) GO TO 6008
C
C*** MODIFY THE SOLUTION VECTOR ACCORDING TO THE MAIN PROGRAM'S
C***      PRINTING FORMAT!!!!
C
   DO   6005   ITOV = 1 ,NTOV
   RRSS (ITOV)=RSOLN(ITOV)
   RSOLN(ITOV)=0.0
 6005                       CONTINUE
```

```
            LTOV = NTOV-1
   DO   6006   ITOV = 1  ,LTOV ,2
            JTOV =(ITOV+  1)/2
   RSOLN(JTOV)=RRSS(ITOV)
 6006                            CONTINUE
   DO   6007   ITOV = 2  ,NTOV ,2
            JTOV =(NTOV+ITOV)/2
   RSOLN(JTOV)=RRSS(ITOV)
 6007                            CONTINUE
C
 6008                            CONTINUE
C
   RETURN
   END
C
C ********************************************************************
C
   SUBROUTINE MODIFY
   1  (NOP ,NE  ,NCN ,IBC ,JBC ,RBC ,NBC ,VBC ,NCOD ,NTOV ,
   2   MAXEL,MAXDF,MAXST,MAXBC,NDF ,MAXBN)
C
   IMPLICIT DOUBLE PRECISION(A-H,O-Z)
C
C FUNCTION
C --------
C   MODIFIES ELEMENT CONNECTIVITY AND ADDRESSING OF THE
C    POINTS TO MAKE THESE CONSISTANT WITH THE PROGRAM
C
   DIMENSION NOP  (MAXEL,MAXST)
   DIMENSION IBC  (MAXBC)
   DIMENSION RBC  (MAXDF)
   DIMENSION JBC  (MAXBC)
   DIMENSION VBC  (MAXBC)
   DIMENSION NCOD (MAXDF)
   DIMENSION INDX (9  )
C
C *** MODIFY ELEMENT CONNECTIVITY
C
   DO 6000 MM =1,NE
   DO 5000 ICN=1,NCN
   INDX(ICN)=NOP(MM,ICN)
 5000 CONTINUE
   DO 5001 ICN=1,NCN
   KCN=INDX(ICN)
   DO 5002 IDF=1,NDF
   IND=NCN*(IDF-1)+ICN
   JND=NDF*(KCN-1)+IDF
   NOP(MM,IND)=JND
 5002 CONTINUE
 5001 CONTINUE
 6000 CONTINUE
C
   IF(NDF .EQ. 1) GO TO 8000
C
C ***  MODIFY ARRAY FOR ADDRESSING BOUNDARY DATA
C
   DO 5999 INP=1,NTOV
   RBC (INP)=0
   NCOD(INP)=0
```

```
 5999 CONTINUE
      DO 6001 INP=1,NBC
      IF(IBC(INP).EQ.0.OR.JBC(INP).EQ.3) GO TO 6001
      ICOD=NDF*(IBC(INP)-1)+JBC(INP)
      NCOD(ICOD)=1
      RBC (ICOD)=VBC(INP)
 6001 CONTINUE
C
 8000 RETURN
      END
C
C*************************************************************************
C
      SUBROUTINE GETNOD (NNP ,CORD ,IDV1 ,IDV2 ,MAXNP,NDIM)
C                 ====
      IMPLICIT DOUBLE PRECISION(A-H,O-Z)
C
C   IDV1 INPUT DEVICE ID
C   IDV2 OUTPUT DEVICE ID
C
      DIMENSION CORD(MAXNP, NDIM)
C
      READ (IDV1,1000) (INP ,(CORD(INP,IDF),IDF=1,2) ,JNP=1,NNP)
      WRITE(IDV2,2000)
      WRITE(IDV2,2010) (JNP ,(CORD(JNP,IDF),IDF=1,2) ,JNP=1,NNP)
C
      RETURN
C
 1000 FORMAT(I5,2E15.8)
 2000 FORMAT('1',///' ',20('*'),' NODAL COORDINATES ',20('*'),//
     1' ',2(7X,'ID.',7X,'X-COORD',7X,'Y-COORD',20X)/)
 2010 FORMAT(' ',I10,2G15.5,20X,I10,2G15.5)
C
      END
C
C ***********************************************************************
C
      SUBROUTINE GETELM (NEL ,NCN ,NODE ,IDV1 ,IDV2 ,MAXEL)
C
      IMPLICIT DOUBLE PRECISION(A-H,O-Z)
C
      DIMENSION NODE (MAXEL, 18)
C
      DO 5000 IEL = 1 ,NEL
 5000 READ (IDV1,1000) MEL ,(NODE(IEL,ICN),ICN=1,NCN)
      WRITE(IDV2,2000)
         DO 5010 JEL = 1 ,NEL
 5010 WRITE(IDV2,2010) JEL ,(NODE(JEL,ICN),ICN=1,NCN)
C
      RETURN
C
 1000 FORMAT(10I5)
 2000 FORMAT('1',///,' ',20('*'),' ELEMENT CONNECTIVITY ',20('*'),//
     1' ',7X,'ID.',5X,'N O D A L -- P O I N T E N T R I E S',/)
 2010 FORMAT(' ',I10,5X,10I8/,' ',15X,10I8/,' ',15X,10I8)
C
      END
C
C ***********************************************************************
```

```
C
   SUBROUTINE GETBCD (NBC,IBC,JBC,VBC,IDV1,IDV2,MAXBC)
C
   IMPLICIT DOUBLE PRECISION(A-H,O-Z)
C

   DIMENSION IBC (MAXBC) ,JBC (MAXBC),VBC (MAXBC)
C
   READ (IDV1,1000) (IBC(IND) ,JBC(IND) ,VBC(IND) ,IND=1,NBC)
   WRITE(IDV2,2000)
   WRITE(IDV2,2010) (IBC(IND) ,JBC(IND) ,VBC(IND) ,IND=1,NBC)
C
   RETURN
C
 1000 FORMAT(2I5,F10.0)
 2000 FORMAT(' ',///,' ',20('*'),' BOUNDARY CONDITIONS ',20('*'),//
   1' ',2(7X,'ID.',2X,'DOF',10X,'VALUE',10X)/)
 2010 FORMAT(' ',5X,2I5,G15.5,15X,2I5,G15.5)
C
   END
C
C ************************************************************************
C
   SUBROUTINE PUTBCV
   1   (NNP ,NBC ,IBC ,JBC ,VBC ,NCOD ,MAXBC,MAXDF,BC)
C
   IMPLICIT DOUBLE PRECISION(A-H,O-Z)
C
C   NCOD ARRAY FOR IDENTIFICATION OF BOUNDARY NODES
C   BC   ARRAY FOR STORING BOUNDARY CONDITION VALUES
C
   DIMENSION IBC (MAXBC) ,JBC (MAXBC) ,VBC (MAXBC)
   DIMENSION NCOD (MAXDF) ,BC (MAXDF)
C
   DO 5000 IND = 1 ,NBC
   IF(JBC(IND).GT.2) GO TO 5000
   JND     = IBC(IND)+(JBC(IND)-1)*NNP
   BC  (JND) = VBC (IND)
   NCOD (JND) = 1
 5000 CONTINUE
C
   RETURN
   END
C
C ************************************************************************
C
   SUBROUTINE PUTBCT
   1   (NBC ,IBC ,JBC ,VBC ,NCOD ,BC  ,MAXBC,MAXDF)
C
   IMPLICIT DOUBLE PRECISION(A-H,O-Z)
C
   DIMENSION IBC (MAXBC) ,JBC (MAXBC) ,VBC (MAXBC)
   DIMENSION NCOD (MAXDF) ,BC (MAXDF)
C
   DO 5000 IND = 1 ,NBC
   IF(JBC(IND).NE.3) GO TO 5000
   JND     = IBC(IND)
   BC   (JND)= VBC(IND)
   NCOD  (JND)= 1
```

```
 5000 CONTINUE
C
    RETURN
    END
C
C ********************************************************************
C
    SUBROUTINE CLEAN
    1  (R1  ,BC  ,NCOD ,NTOV ,MAXDF, MAXBN)
C
    IMPLICIT DOUBLE PRECISION(A-H,O-Z)
C
    DIMENSION R1  (MAXDF)
    DIMENSION BC  (MAXDF)
    DIMENSION NCOD (MAXDF)
    COMMON/ONE/ STIFF(2000,300)
C
C FUNCTION
C --------
C   CLEANS THE USED ARRAYS AND PREPARES THEM FOR REUSE
C
       DO 5000 I  = 1,NTOV
          R1(I) = 0.0
          BC(I) = 0.0
          NCOD(I)= 0
 5000         CONTINUE
       DO 5020  I = 1,600
       DO 5020  J = 1,500
        STIFF(J,I)= 0.0
 5020         CONTINUE
C
    RETURN
    END
C
C ********************************************************************
C
    SUBROUTINE SETPRM
    1  (NNP ,NEL   ,NCN ,NODE ,NDF ,MAXEL,MAXST)
C
    IMPLICIT DOUBLE PRECISION(A-H,O-Z)
C
    DIMENSION NODE (MAXEL,MAXST)
C
C FUNCTION
C --------
C SETS THE LOCATION DATA FOR NODAL DEGREES OF FREEDOM
C
    DO 5000 IEL = 1 ,NEL
    DO 5000 ICN = 1 ,NCN
        KCN =NODE(IEL,ICN)
        JCN =ICN+(NDF-1)*NCN
        LCN =KCN+(NDF-1)*NNP
C
    NODE(IEL,JCN) = LCN
C
 5000 CONTINUE
C
    RETURN
    END
```

```
C
C  ***********************************************************************
C
   SUBROUTINE GETMAT (NEL,PMAT,IDV1 ,IDV2 ,MAXEL,RTEM)
C
   IMPLICIT DOUBLE PRECISION(A-H,O-Z)
C
   DIMENSION PMAT (MAXEL,  8)
C
   WRITE(IDV2,2000)
C
   READ (IDV1,1000) RVISC ,POWER ,TREF ,TBCO ,RODEN ,CP ,CONDK
C
   READ (IDV1,1010) RPLAM
                IFROM = 1
                ITO  = NEL
        IF(TREF .EQ.0.) TREF = 0.001
   DO 5010 IEL = IFROM ,ITO
     PMAT(IEL,1) = RVISC
     PMAT(IEL,2) = RPLAM
     PMAT(IEL,3) = POWER
     PMAT(IEL,4) = TREF
     PMAT(IEL,5) = TBCO
     PMAT(IEL,6) = RODEN
     PMAT(IEL,7) = CP
     PMAT(IEL,8) = CONDK
     RTEM    = TREF
C
C *** PARAMETERS OF THE POWER-LAW MODEL
C
C *** RVISC = MEU NOUGHT; CONSISTENCY COEFFICIENT
C *** RPLAM = PENALTY PARAMETER
C *** POWER = POWER LAW INDEX
C *** TREF = REFERENCE TEMPERATURE
C *** TBCO = COEFFICIENT b IN THE POWER LAW MODEL
C
C *** PHYSICAL PARAMETERS
C
C *** RODEN = MATERIAL DENSITY
C *** CP  = SPECIFIC HEAT
C *** CONDK = HEAT CONDUCTIVITY COEFFICIENT
C
 5010 CONTINUE
C
   WRITE(IDV2,2010) IFROM ,ITO ,RVISC ,RPLAM ,POWER
   WRITE(IDV2,2020)
   WRITE(IDV2,2030) TREF ,TBCO
   WRITE(IDV2,2040)
   WRITE(IDV2,2050) RODEN ,CP  ,CONDK
 5000 CONTINUE
C
   RETURN
C
 1000 FORMAT(7F10.0)



 1010 FORMAT(F15.0)
```

```
2000 FORMAT('0',//' ',35('*'),'MATERIAL PROPERTIES',35('*'),//
   1' ',7X,'ID.',5X,'EID.(FROM-TO)',7X,'CONSISTENCY COEFFICIENT',8X,
   2'PENALTY PARAMETER',8X,'POWER LAW INDEX',/)
2010 FORMAT(' ',I10,I14,I4,3G20.5)
2020 FORMAT(10X,#** REFERENCE TEMPERATURE **  ** COEFFICIENT b **')
2030 FORMAT(17X,G10.3,17X,G10.3)
2040 FORMAT
   1(10X,' ** DENSITY **  ** SPECIFIC HEAT **  ** CONDUCTIVITY ** ')
2050 FORMAT(14X,G10.3,8X,G10.3,11X,G10.3)
C
   END
C
C ********************************************************************
C
   SUBROUTINE VISCA
   1      (RVISC,POWER,VISC,SRATE,STEMP,RTEM,TCO)
C
   IMPLICIT DOUBLE PRECISION(A-H,O-Z)
C
C *** CALCULATE SHEAR-DEPENDENT VISCOSITY
C
   PINDX = (POWER-1.)/2
   VISC  = RVISC*((4.*SRATE)**PINDX)*EXP(-TCO*(STEMP-RTEM))
C
   RETURN
   END
C
C ********************************************************************
C
   SUBROUTINE CONTOL
   1(VEL,TEMP,ITER,NTOV,NNP,MAXNP,MAXDF,ERROV,ERROT,VET,TET)
C
   IMPLICIT DOUBLE PRECISION(A-H,O-Z)
C
   DIMENSION VEL (MAXDF),TEMP (MAXNP)
   DIMENSION VET (MAXDF),TET (MAXNP)
C
C *** CALCULATE DIFFERENCE BETWEEN VELOCITIES IN CONSECUTIVE ITERATIONS
C
   ERRV = 0.0
   TORV = 0.0
   ERRT = 0.0
   TORT = 0.0
         DO 1000 ICHECK = 1,NTOV
   IF(ITER.EQ.1) VET(ICHECK) = 0.0
         ERRV = ERRV + (VEL(ICHECK)-VET(ICHECK))**2
         TORV = TORV + (VEL(ICHECK))**2
C
   VET(ICHECK) = VEL(ICHECK)
C
 1000       CONTINUE
         ERROV= ERRV/TORV
C
C *** CALCULATE DIFFERENCE BETWEEN TEMPERATURES IN CONSECUTIVE
C         ITERATIONS
         DO 2000 ICHECK = 1,NNP
   IF(ITER.EQ.1) TET(ICHECK) = 0.0
         ERRT = ERRT + (TEMP(ICHECK)-TET(ICHECK))**2
         TORT = TORT + (TEMP(ICHECK))**2
```

```
C
   TET(ICHECK) = TEMP(ICHECK)
C
 2000        CONTINUE
          ERROT= ERRT/TORT
C
   RETURN
   END
C
C  **********************************************************************
C
   SUBROUTINE OUTPUT
   1   (NNP ,VEL ,TEMP ,MAXDF,MAXNP)
C
   IMPLICIT DOUBLE PRECISION(A-H,O-Z)
C
   DIMENSION VEL (MAXDF),TEMP (MAXNP)
C
   WRITE(60,6000)
 5999 FORMAT(' ID.   UX        UY        T'/)
C
 6000 FORMAT('*** NODAL VELOCITIES AND NODAL TEMPERATURES ***'/)
          DO 6001 INP = 1,NNP
              JNP = INP + NNP
   WRITE(60,6002)
   1        INP ,VEL(INP),VEL(JNP) ,TEMP(INP)
 6002 FORMAT(I5,2E13.4,F13.4)
 6001 CONTINUE
C
   RETURN
   END
C
C  **********************************************************************
C
```

+-----------------------+
| STORE OUTPUT |
| FILES FOR |
| POST-PROCESSING |
+-----------------------+

REFERENCES

Gerald, C. F. and Wheatley, P. O., 1984. *Applied Numerical Analysis*, 3rd edn, Addison-Wesley, Reading, MA.

Lapidus, L. and Pinder, G. F., 1982. *Numerical Solution of Partial Differential Equations in Science and Engineering*, Wiley, New York.

8

Appendix – Summary of Vector and Tensor Analysis

Commonly Used Coordinate Systems, Basic Definitions and Operations

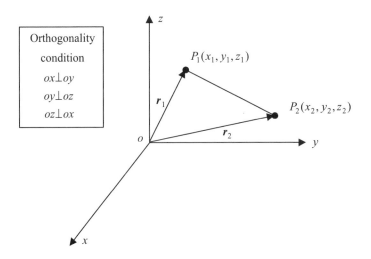

Figure A.1

In a Cartesian (planar) coordinate system of (x, y, z) shown in Appendix Figure 1 the position vectors for points P_1 and P_2 are given as

$$r_1 = x_1 i + y_1 j + z_1 k \quad \text{and} \quad r_2 = x_2 i + y_2 j + z_2 k$$

where i, j and k are unit vectors in the x, y and z directions, respectively. The magnitude of line $P_1 P_2$ (i.e. the distance between points P_1 and P_2) is thus found as

$$d = \overline{P_1 P_2} = \sqrt{(x_2 - x_1)^2 + (y_2 - y_1)^2 + (z_2 - z_1)^2}$$

The direction cosines of the vector $\boldsymbol{r} = x\boldsymbol{i} + y\boldsymbol{j} + z\boldsymbol{k}$ are the numbers defined as

$$\begin{cases} \cos \alpha = \dfrac{x}{|\boldsymbol{r}|} \\[2mm] \cos \beta = \dfrac{y}{|\boldsymbol{r}|} \\[2mm] \cos \gamma = \dfrac{z}{|\boldsymbol{r}|} \end{cases} \qquad [\cos^2\alpha + \cos^2\beta + \cos^2\gamma = 1]$$

where α, β and γ are the angles which the vector makes with the positive directions of the coordinate axis shown in Appendix Figure 1 and $|\boldsymbol{r}|$ is the vector magnitude found as $|\boldsymbol{r}| = \sqrt{x^2 + y^2 + z^2}$. Other commonly used orthogonal coordinate systems are cylindrical (polar) and spherical systems shown as

Cylindrical (r, θ, z)

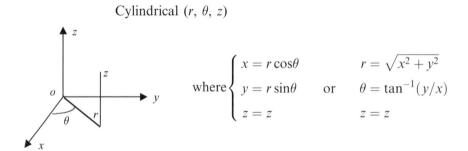

$$\text{where} \begin{cases} x = r\cos\theta \\ y = r\sin\theta \\ z = z \end{cases} \quad \text{or} \quad \begin{aligned} r &= \sqrt{x^2 + y^2} \\ \theta &= \tan^{-1}(y/x) \\ z &= z \end{aligned}$$

Figure A.2

Spherical (r, θ, ϕ)

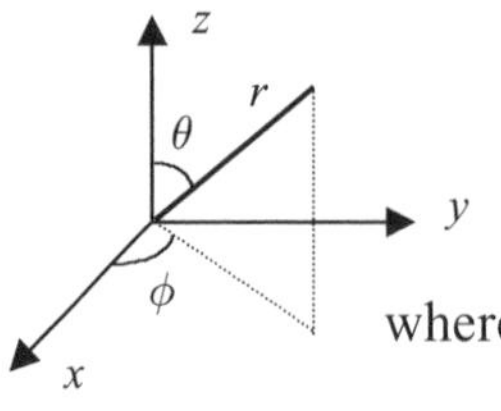

$$\text{where} \begin{cases} x = r\sin\theta \cos\phi \\ y = r\sin\theta \sin\phi \\ z = r\cos\theta \end{cases} \quad \text{or} \quad \begin{aligned} r &= \sqrt{x^2 + y^2 + z^2} \\ \phi &= \tan^{-1}(y/x) \\ \theta &= \cos^{-1}\left(z/\sqrt{x^2 + y^2 + z^2}\right) \end{aligned}$$

Figure A.3

Orthogonal transformation of a Cartesian vector A with components A_1, A_2 and A_3 in the system of $o123$, under rotation of the coordinate system to $o\overline{1}\,\overline{2}\,\overline{3}$ is expressed by the following equation

$\bar{A}_J = d_{JI} A_I$ (summation convention over the repeated index I is implied)

where $\bar{A}_J$ denote the components of the vector in coordinate system $o\bar{1}\,\bar{2}\,\bar{3}$ and d_{JI} are the cosines of the angles between the old axis oi ($i = 1,2,3$) and the new one $o\bar{j}$. Therefore

$$\begin{Bmatrix} \bar{A}_1 \\ \bar{A}_2 \\ \bar{A}_3 \end{Bmatrix} = \begin{Bmatrix} d_{11} & d_{12} & d_{13} \\ d_{21} & d_{22} & d_{23} \\ d_{31} & d_{32} & d_{33} \end{Bmatrix} \begin{Bmatrix} A_1 \\ A_2 \\ A_3 \end{Bmatrix}$$

A vector which remains unchanged in such a transformation (i.e. $\bar{A} = A$) is said to be invariant.

8.1   VECTOR ALGEBRA

- Vectors A and B are equal if they have the same magnitude and direction regardless of the position of their origin.
- A vector whose magnitude is equal to the magnitude of A but is in the opposite direction is denoted by $-A$.
- The sum (resultant) of vectors A and B is a vector C found using the parallelogram law graphically shown as

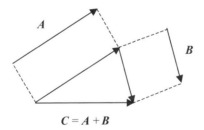

$$C = A + B$$

Figure A.4

- The difference of A and B is a vector C given as $C = A - B = A + (-B)$. For $A = B$ the resultant vector C is a zero vector (i.e. its magnitude is zero and has no specific direction).
- The product of A by a scalar m is a vector having the same direction as A with a magnitude of $m\bar{A}$.
- $A + B = B + A$ (commutative law for addition)
- $A + (B + C) = (A + B) + C$ (associative law for addition)
- $mA = Am$ (commutative law for multiplication)
- $m(nA) = (mn)A$ (associative law for multiplication)
- $(m + n)\,A = mA + nA$ (distributive law)
- $m(A + B) = mA + mB$ (distributive law)

- The scalar (dot) product of two vectors is a number found as

  $A.B = |A||B| \cos \theta$ where $0 < \theta < \pi$ is the angle between the directions of A and B.

  Alternatively in terms of components of A and B, $A.B = A_1B_1 + A_2B_2 + A_3B_3$.

  If $A.B = 0$ and $A, B \neq 0$ then $A \perp B$. If $A.B = 1$ then A and B are parallel (specifically $i.i = j.j = k.k = 1,$ $i.j = j.k = k.i = 0$).

  $A.B = B.A$ (commutative law for scalar product)

  $A.(B + C) = A.B + A.C$ (distributive law for scalar product)

  $m(A.B) = (mA).B = A.(mB) = (A.B)m$

- Vector (cross) product of vectors A and B is a vector C shown as $C = A \times B$ with a magnitude of $|C| = |A||B| \sin \theta$ where $0 < \theta < \pi$ is the angle between the directions of A and B. The direction of C is perpendicular to the plane of A and B such that A, B and C make a right-handed system shown as

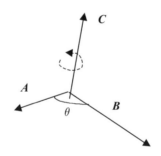

Figure A.5

Alternatively in terms of the components of A and B, the vector product $A \times B$ can be expressed as the following determinant

$$A \times B = \begin{vmatrix} i & j & k \\ A_1 & A_2 & A_3 \\ B_1 & B_2 & B_3 \end{vmatrix}$$

If $A \times B = 0$ then A and B are parallel

(specifically $i \times i = j \times j = k \times k = 0,$ $i \times j = k, j \times k = i, k \times i = j$).

$A \times B = -B \times A$

$A \times (B + C) = A \times B + A \times C$ (distributive law)

8.2 SOME VECTOR CALCULUS RELATIONS

If $A = A(x, y, z)$ then $dA = \dfrac{\partial A}{\partial x} dx + \dfrac{\partial A}{\partial y} dy + \dfrac{\partial A}{\partial z} dz$

Let P be a vector whose components are functions of a scalar variable (e.g. time-dependent position vector of a point P in a three-dimensional domain)

$$P(t) = x(t)i + y(t)j + z(t)k$$

then

$$\frac{dP}{dt} = \frac{dx}{dt}i + \frac{dy}{dt}j + \frac{dz}{dt}k$$

If A, B and C are differentiable vector functions of scalar t and ϕ is a differentiable function of t then

$$\frac{d}{dt}(A + B) = \frac{dA}{dt} + \frac{dB}{dt}$$

$$\frac{d}{dt}(\phi A) = \phi \frac{dA}{dt} + \frac{d\phi}{dt}A$$

$$\frac{d}{dt}(A.B) = A \cdot \frac{dB}{dt} + \frac{dA}{dt} \cdot B \quad \text{and} \quad \frac{\partial}{\partial t}(A.B) = A \cdot \frac{\partial B}{\partial t} + \frac{\partial A}{\partial t} \cdot B$$

$$\frac{d}{dt}(A \times B) = A \times \frac{dB}{dt} + \frac{dA}{dt} \times B \quad \text{and} \quad \frac{\partial}{\partial t}(A \times B) = A \times \frac{\partial B}{\partial t} + \frac{\partial A}{\partial t} \times B$$

The vector differential operator *del* (or *nabla*) written as ∇ is defined by

$$\nabla \equiv \frac{\partial}{\partial x}i + \frac{\partial}{\partial y}j + \frac{\partial}{\partial z}k \equiv i\frac{\partial}{\partial x} + j\frac{\partial}{\partial y} + k\frac{\partial}{\partial z}$$

where i, j and k are the unit vectors in a Cartesian coordinate system. The del operator has properties analogous to those of ordinary vectors.

The *gradient* of a scalar $\phi(x, y, z)$ is defined by

$$\nabla(\phi) = \frac{\partial \phi}{\partial x}i + \frac{\partial \phi}{\partial y}j + \frac{\partial \phi}{\partial z}k$$

The *divergence* of a vector $V(x, y, z)$ is defined by

$$\nabla . V = \left(\frac{\partial}{\partial x}\boldsymbol{i} + \frac{\partial}{\partial y}\boldsymbol{j} + \frac{\partial}{\partial z}\boldsymbol{k}\right) \cdot (V_1\boldsymbol{i} + V_2\boldsymbol{j} + V_3\boldsymbol{k}) = \frac{\partial V_1}{\partial x} + \frac{\partial V_2}{\partial y} + \frac{\partial V_3}{\partial z}$$

(note that $\nabla . V \neq V . \nabla$).

The *curl* of a vector $V(x, y, z)$ is defined by

$$\nabla \times V = \left(\frac{\partial}{\partial x}\boldsymbol{i} + \frac{\partial}{\partial y}\boldsymbol{j} + \frac{\partial}{\partial z}\boldsymbol{k}\right) \times (V_1\boldsymbol{i} + V_2\boldsymbol{j} + V_3\boldsymbol{k}) = \begin{vmatrix} \boldsymbol{i} & \boldsymbol{j} & \boldsymbol{k} \\ \frac{\partial}{\partial x} & \frac{\partial}{\partial y} & \frac{\partial}{\partial z} \\ V_1 & V_2 & V_3 \end{vmatrix}$$

$$\nabla \times (\nabla \times A) = \nabla(\nabla . A) - \nabla^2 A$$

where

$$\nabla^2 = \frac{\partial^2}{\partial x^2} + \frac{\partial^2}{\partial y^2} + \frac{\partial^2}{\partial z^2}$$

(in a Cartesian system) is called the *Lapacian* operator.

Let $V(t) = V_x(t)\boldsymbol{i} + V_y(t)\boldsymbol{j} + V_z(t)\boldsymbol{k}$, then

$$\int V(t)\mathrm{d}t = \boldsymbol{i} \int V_x(t)\mathrm{d}t + \boldsymbol{j} \int V_y(t)\mathrm{d}t + \boldsymbol{k} \int V_z(t)\mathrm{d}t$$

If there exists a vector $R(t)$ where

$$V(t) = \frac{\mathrm{d}}{\mathrm{d}t}[R(t)]$$

then

$$\int V(t)\mathrm{d}t = R(t) + c$$

where c is an arbitrary constant vector.

8.2.1 Divergence (Gauss) theorem

If V is the volume bounded by a closed surface S and A is a vector function of position with continuous derivatives, then

$$\iiint_V \nabla.A \, dV = \iint_S A.n \, dS = \oiint_S A.dS$$

Note that the surface integral $\oiint_S A.dS$ denotes the flux of A over the closed surface S.

8.2.2   Stokes theorem

Let S be an open, two-sided surface bounded by a curve C, then the line integral of vector A (curve C is traversed in the positive direction) is expressed as

$$\oint_C A.dr = \iint_S (\nabla \times A).n \, dS = \iint_S (\nabla \times A).dS$$

Note that Green's theorem in the plane expressed as

$$\oint_L [f_1(x,y)dx + f_2(x,y)dy] = \iint_R \left(\frac{\partial f_2}{\partial x} - \frac{\partial f_1}{\partial y}\right) dxdy$$

is the special case of the Stokes theorem. It should also be noted that the Gauss' divergence theorem can be obtained by generalization of Green's theorem in the plane by replacing the region R and its boundary curve C with a space region and its closing surface.

8.2.3   Reynolds transport theorem

This theorem provides a convenient means for obtaining rate of change of a vector field function over a volume $V(t)$ as

$$\frac{D}{Dt} \iiint_{V(t)} F(r,t)dV = \iiint_{V(t)} \left[\frac{\partial F}{\partial t} + \nabla \cdot (Fv)\right] dV$$

where

$$\frac{D}{Dt} = \frac{\partial}{\partial t} + v.\nabla$$

is the material (substantial) time derivative and v is the velocity vector.

8.2.4   Covariant and contravariant vectors

These definitions arise from the transformation properties of vectors and can be summarized as follows: If in the transformation of the coordinate system $(x^1, x^2, \ldots, x^n)$ to another system $(\bar{x}^1, \bar{x}^2, \ldots, \bar{x}^n)$ quantities $A_1, A_2, \ldots A_n$ transform to $\bar{A}_1, \bar{A}_2, \ldots, \bar{A}_n$, such that

$$\bar{A}_p = \sum_{q=1}^{n} \frac{\partial x^q}{\partial \bar{x}^p} A_q, p = 1, 2, \ldots, n$$

then $A_1, A_2, \ldots, A_n$ are the components of a covariant vector. Similarly $A^1, A^2, \ldots, A^n$ are said to be components of a contravariant vector if in the transformation of the coordinate system $(x^1, x^2, \ldots, x^n)$ to another system $(\bar{x}^1, \bar{x}^2, \ldots, \bar{x}^n)$ they transform according to

$$\bar{A}^p = \sum_{q=1}^{n} \frac{\partial \bar{x}^p}{\partial x^q} A^q, p = 1, 2, \ldots, n$$

8.2.5   Second order tensors

Field variables identified by their magnitude and two associated directions are called second-order tensors (by analogy a scalar is said to be a zero-order tensor and a vector is a first-order tensor). An important example of a second-order tensor is the physical function stress which is a surface force identified by magnitude, direction and orientation of the surface upon which it is acting. Using a mathematical approach a second-order Cartesian tensor is defined as an entity having nine components T_{ij}, i, j = 1, 2, 3, in the Cartesian coordinate system of $o123$ which on rotation of the system to $o\bar{1}\,\bar{2}\,\bar{3}$ become

$$\bar{T}_{pq} = d_{ip} d_{jq} T_{ij}$$

where d_{ip} and d_{jq} are the cosines of the angles between the new and old coordinate axis. By the orthogonality conditions of these direction cosines the inverse of this transformation is expressed as

$$T_{ij} = d_{ip} d_{jq} \bar{T}_{pq}$$

Components of a second-order tensor T in a three-dimensional frame of reference are written as the following 3 × 3 matrix

$$\begin{bmatrix} T_{11} & T_{12} & T_{13} \\ T_{21} & T_{22} & T_{23} \\ T_{31} & T_{32} & T_{33} \end{bmatrix}$$

It follows that by using component forms second-order tensors can also be manipulated by rules of matrix analysis.

A second-order tensor whose components satisfy $T_{ij} = T_{ji}$ is called symmetric and has six distinct components. If $T_{ij} = -T_{ji}$ then the tensor is said to be antisymmetric.

To obtain the transpose of a tensor T the indices of its components (originally given as T_{pq}) are transposed such that

$$T^{\mathrm{T}} = \sum_p \sum_q \alpha_p \alpha_q \, T_{qp}$$

where α_p is the unit vector in direction p. Tensors of second rank are shown either using expanded or Dyadic notations. Dyadic forms are generalization of vectors and are shown as two vectors together without brackets or multiplication symbols. For example AB denotes the dyadic product of vectors A and B which is a second-order tensor (a further generalization leading to 'triads' denote third-order tensors). In terms of its components AB is shown in the expanded form as

$$AB = \sum_p \sum_q \alpha_p \alpha_q \, A_i B_j$$

Note that $AB \neq BA$ and $(AB)^{\mathrm{T}} = BA$.

The dyadic products formed by unit vectors i, j and k (i.e. ii, jj, ij, ji, etc.) are called unit dyads (note that $ij \neq ji$) and represent ordered pairs of coordinate directions.

An isotropic tensor is one whose components are unchanged by rotation of the coordinate system.

8.3 TENSOR ALGEBRA

The unit second-order tensor is the Kronecker delta, δ, whose components are given as the following matrix

$$\delta_{ij} = \begin{vmatrix} 1 & 0 & 0 \\ 0 & 1 & 0 \\ 0 & 0 & 1 \end{vmatrix} \qquad \text{i.e. } \delta_{ij} = \begin{Bmatrix} 0 \text{ if } i \neq j \\ 1 \text{ if } i = j \end{Bmatrix}$$

- The sum of two tensors is found by adding their corresponding components as

$$T + D = \sum_p \sum_q \alpha_p \alpha_q (T_{pq} + D_{pq})$$

- The scalar (double-dot) product of two tensors is found as follows

$$\boldsymbol{T} : \boldsymbol{D} = \left(\left[\sum_p \sum_q \alpha_p \alpha_q \, T_{pq} \right] : \left[\sum_k \sum_l \alpha_k \alpha_l \, D_{kl} \right] \right)$$

$$= \sum_p \sum_q \sum_k \sum_l (\alpha_p \alpha_q : \alpha_k \alpha_l) T_{pq} D_{kl}$$

$$= \sum_p \sum_q \sum_k \sum_l \delta_{pl} \delta_{qk} T_{pq} D_{kl}$$

$$= \sum_p \sum_q T_{pq} D_{qp}$$

Or using dyadic products

$$(\boldsymbol{rs} : \boldsymbol{td}) = \sum_p \sum_q r_p s_q t_q d_p$$

- The tensor (single-dot) product of two tensors is found as follows

$$\boldsymbol{T} \cdot \boldsymbol{D} = \left\{ \left[\sum_p \sum_q \alpha_p \alpha_q \, T_{pq} \right] \cdot \left[\sum_k \sum_l \alpha_k \alpha_l \, D_{kl} \right] \right\}$$

$$= \sum_p \sum_q \sum_k \sum_l (\alpha_p \alpha_q \cdot \alpha_k \alpha_l) T_{pq} D_{kl}$$

$$= \sum_p \sum_q \sum_k \sum_l \delta_{qk} \alpha_p \alpha_l T_{pq} D_{kl}$$

$$= \sum_p \sum_l \alpha_p \alpha_l \left(\sum_q T_{pq} D_{ql} \right) \qquad \text{i.e. } pl \text{ component} \atop \text{of the product is} \quad \sum_q T_{pq} D_{ql}$$

- The vector (dot) product of a tensor with a vector is found as follows

$$\boldsymbol{T} \cdot \boldsymbol{A} = \left\{ \left[\sum_p \sum_q \alpha_p \alpha_q \, T_{pq} \right] \cdot \left[\sum_k \alpha_k \, A_k \right] \right\}$$

$$= \sum_p \sum_q \sum_k (\alpha_p \alpha_q \cdot \alpha_k) T_{pq} A_k$$

$$= \sum_p \sum_q \sum_k \alpha_p \delta_{qk} T_{pq} A_k$$

$$= \sum_p \alpha_p \left(\sum_q T_{pq} A_q \right) \qquad \text{i.e. } p\text{th component} \atop \text{of the product is} \quad \sum_q T_{pq} A_q$$

- The tensor (cross) product of a tensor with a vector is found as follows

$$T \times A = \left\{ \left[\sum_p \sum_q \alpha_p \alpha_q \, T_{pq} \right] \times \left[\sum_k \alpha_k A_k \right] \right\}$$

$$= \sum_p \sum_q \sum_k (\alpha_p \alpha_q \times \alpha_k) T_{pq} A_k$$

$$= \sum_p \sum_q \sum_k \sum_l \varepsilon_{qkl} \, \alpha_p \alpha_l \, T_{pq} A_k$$

$$= \sum_p \sum_l \alpha_p \alpha_l \left(\sum_q \sum_k \varepsilon_{qkl} T_{pq} A_k \right)$$

i.e. pl component of the product is

$$\sum_q \sum_k \varepsilon_{qkl} T_{pq} A_k$$

where ε_{ijk} is the permutation symbol which is a third-order tensor defined as

$$\varepsilon_{ijk} = \begin{cases} 0, \text{ if any two of } i, j, k \text{ are the same} \\ 1, \text{ if } ijk \text{ is an even permutation of } 1,2,3 \\ -1, \text{ if } ijk \text{ is an odd permutation of } 1,2,3 \end{cases}$$

- The magnitude of a tensor is defined as

$$|T| = T = \sqrt{\tfrac{1}{2}(T : T^{\mathrm{T}})} = \sqrt{\tfrac{1}{2} \sum_i \sum_j T_{ij}^2}$$

- Any tensor may be represented as the sum of a symmetric part and an antisymmetric part

$$T_{ij} = \tfrac{1}{2}[T_{ij} + T_{ji}] + \tfrac{1}{2}[T_{ij} - T_{ji}]$$

The operation of identifying two indices of a tensor and so summing on them is known as contraction, $T_{ij} = T_{11} + T_{22} + T_{33}$.

8.3.1 Invariants of a second-order tensor (T)

The following three scalars remain independent of the choice of coordinate system in which the components of T are defined and hence are called the invariants of tensor T:

- The first invariant is the trace of the tensor, found as

$$I = \text{trace of } T = \text{tr } T = \sum_i T_{ii} \equiv T_{11} + T_{22} + T_{33}$$

(sum of diagonal terms in the components matrix)

- The second invariant is the trace of T^2 i.e. $\{T.T\}$, found as

$$\text{II} = \text{trace of } T^2 = \text{tr } T^2 = \sum_i \sum_j T_{ij}T_{ji}$$

It can hence be seen that the magnitude of a symmetric tensor S is related to its second invariant as

$$|S| = \sqrt{\tfrac{1}{2}\sum_i \sum_j S_{ij}^2} = \sqrt{\tfrac{1}{2}\sum_i \sum_j S_{ij}S_{ij}} = \sqrt{\tfrac{1}{2}\sum_i \sum_j S_{ij}S_{ji}} = \sqrt{\tfrac{1}{2}\text{II}}$$

- The third invariant is the trace of T^3, i.e. $\{T.T^2\}$, found as

$$\text{III} = \text{trace of } T^3 = \text{tr } T^3 = \sum_i \sum_j \sum_k T_{ij}T_{jk}T_{ki}$$

8.4 SOME TENSOR CALCULUS RELATIONS

$$[\nabla \cdot rs] = [r \cdot \nabla s] + s(\nabla \cdot r)$$

$$(a\boldsymbol{\delta} : \nabla v) = a(\nabla \cdot v)$$

$$(\nabla \cdot a\boldsymbol{\delta}) = \nabla a$$

$$(\nabla \cdot aT) = (\nabla a \cdot T) + a(\nabla.T)$$

Analogous to vector operations the tensorial form of the divergence theorem is written as

$$\iiint_V \frac{\partial T_{ij}}{\partial x_i}\,dV = \iint_S T_{ij}n_i\,ds$$

Analogous to vector operations the tensorial form of Stokes theorem is written as

$$\iint_S (\nabla.T).n\,ds = \oint_c T.ds$$

8.4.1 Covariant, contravariant and mixed tensors

Similar to vectors, based on the transformation properties of the second tensors the following three types of covariant, contravariant and mixed components are defined

$$
\begin{cases}
\bar{A}_{pr} = \dfrac{\partial x^q}{\partial \bar{x}^p}\dfrac{\partial x^s}{\partial \bar{x}^r} A_{qs} \quad \text{covariant} \\[4mm]
\bar{A}^{pr} = \dfrac{\partial \bar{x}^p}{\partial x^q}\dfrac{\partial \bar{x}^r}{\partial x^s} A^{qs} \quad \text{contravariant} \qquad \text{(summation convention is used)} \\[4mm]
\bar{A}^p_r = \dfrac{\partial \bar{x}^p}{\partial x^q}\dfrac{\partial x^s}{\partial \bar{x}^r} A^q_s \quad \text{mixed}
\end{cases}
$$

Note that convected derivatives of the stress (and rate of strain) tensors appearing in the rheological relationships derived for non-Newtonian fluids will have different forms depending on whether covariant or contravariant components of these tensors are used. For example, the convected time derivatives of covariant and contravariant stress tensors are expressed as

$$
\frac{\mathcal{D}T}{\mathcal{D}t} = \frac{\partial T}{\partial t} + (v.\nabla)T - (\Omega\times T - T\times\Omega) + 2D.T \qquad \text{(covariant tensor } T)
$$

and

$$
\frac{\mathcal{D}T}{\mathcal{D}t} = \frac{\partial T}{\partial t} + (v.\nabla)T - (\Omega\times T - T\times\Omega) - 2D.T \qquad \text{(contravariant tensor } T)
$$

where $\Omega = \frac{1}{2}(\nabla \times v)$ and $D = \frac{1}{2}[\nabla v + (\nabla v)^{\mathrm{T}}]$ are the vorticity vector and rate of deformation tensors in the flow field, respectively. Note that the convected time derivative $\mathcal{D}/\mathcal{D}t$ is the special case of the general time derivative $\Delta_{a,b,c}/\Delta_t$ used in the text. Using the component forms the above time derivatives are written as

$$
\frac{\mathcal{D}T_{ij}}{\mathcal{D}t} = \frac{\partial T_{ij}}{\partial t} + v^k\frac{\partial T_{ij}}{\partial x_k} + \frac{\partial v^k}{\partial x_i}T_{kj} + \frac{\partial v^k}{\partial x_j}T_{ik}
$$

(covariant components corresponding to lower-convected derivative)

and

$$
\frac{\mathcal{D}T^{ij}}{\mathcal{D}t} = \frac{\partial T^{ij}}{\partial t} + v^k\frac{\partial T^{ij}}{\partial x_k} - \frac{\partial v^i}{\partial x_k}T^{kj} - \frac{\partial v^j}{\partial x_k}T^{ik}
$$

[contravariant components corresponding to upper-convected (or co-deformational) derivative]

8.4.2 The length of a line and metric tensor

In a Cartesian coordinate system the differential of arc length of a line is defined as $ds = \sqrt{dx^2 + dy^2 + dz^2}$ (and hence $ds^2 = dx^2 + dy^2 + dz^2$). After transformation from the Cartesian system (x, y, z) to a general three-dimensional curvilinear coordinate system (ξ) this can be written as

$$ds^2 = \sum_{p=1}^{3} \sum_{q=1}^{3} g_{pq}\, d\xi^p d\xi^q$$

or using summation convention

$$ds^2 = g_{pq}\, d\xi^p d\xi^q$$

where the quantities g_{pq} are elements of a matrix found as the dot products of the pairs of basic tangential vectors as

$$g_{pq} = X_{\xi^p} \cdot X_{\xi^q} = \frac{\partial X^k}{\partial \xi^p}\frac{\partial X^k}{\partial \xi^q} \qquad p,q,k = 1,\ldots,n$$

The following figure is a two-dimensional example illustrating the symbols used in the previous relationship.

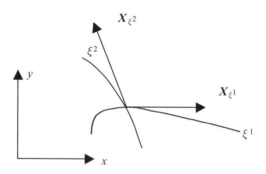

Figure A.6

The matrix g_{pq} represents the components of a covariant second-order tensor called the 'metric tensor', because it defines distance measurement with respect to coordinates $\xi^1,\ldots,\xi^n$. To illustrate the application of this definition in the coordinate transformation we consider the transformation from coordinate system x to X, shown as

$$x \rightarrow X$$

For simplicity we assume a two-dimensional case, therefore

$$x \Rightarrow x_1, x_2$$
$$X \Rightarrow X_1, X_2$$

the Jacobian of the transformation is written as

$$
J = \begin{vmatrix} \dfrac{\partial X_1}{\partial x_1} & \dfrac{\partial X_1}{\partial x_2} \\[3mm] \dfrac{\partial X_2}{\partial x_1} & \dfrac{\partial X_2}{\partial x_2} \end{vmatrix}
$$

The transformation of components of a second-order tensor, given as the following matrix in the coordinate system x

$$
a = \begin{vmatrix} a_{11} & a_{12} \\ a_{21} & a_{22} \end{vmatrix}
$$

is based on

$$
\bar{A}_{pr} = \sum_{s=1}^{2}\sum_{q=1}^{2} \frac{\partial x^q}{\partial \bar{x}^p} \cdot \frac{\partial x^s}{\partial \bar{x}^r} A_{qs}
$$

where

$$
\frac{\partial x^q}{\partial \bar{x}^p} \cdot \frac{\partial x^s}{\partial \bar{x}^r}
$$

represent the components of g_{pq}. Therefore

$$
A_{11} = \frac{\partial x_1}{\partial X_1}\cdot\frac{\partial x_1}{\partial X_1}a_{11} + \frac{\partial x_1}{\partial X_1}\cdot\frac{\partial x_2}{\partial X_1}a_{12} + \frac{\partial x_2}{\partial X_1}\cdot\frac{\partial x_1}{\partial X_1}a_{21} + \frac{\partial x_2}{\partial X_1}\cdot\frac{\partial x_2}{\partial X_1}a_{22}
$$

$$
A_{12} = \frac{\partial x_1}{\partial X_1}\cdot\frac{\partial x_1}{\partial X_2}a_{11} + \frac{\partial x_1}{\partial X_1}\cdot\frac{\partial x_2}{\partial X_2}a_{12} + \frac{\partial x_2}{\partial X_1}\cdot\frac{\partial x_1}{\partial X_2}a_{21} + \frac{\partial x_2}{\partial X_1}\cdot\frac{\partial x_2}{\partial X_2}a_{22}
$$

$$
A_{21} = \frac{\partial x_1}{\partial X_2}\cdot\frac{\partial x_1}{\partial X_1}a_{11} + \frac{\partial x_1}{\partial X_2}\cdot\frac{\partial x_2}{\partial X_1}a_{12} + \frac{\partial x_2}{\partial X_2}\cdot\frac{\partial x_1}{\partial X_1}a_{21} + \frac{\partial x_2}{\partial X_2}\cdot\frac{\partial x_2}{\partial X_1}a_{22}
$$

$$
A_{22} = \frac{\partial x_1}{\partial X_2}\cdot\frac{\partial x_1}{\partial X_2}a_{11} + \frac{\partial x_1}{\partial X_2}\cdot\frac{\partial x_2}{\partial X_2}a_{12} + \frac{\partial x_2}{\partial X_2}\cdot\frac{\partial x_1}{\partial X_2}a_{21} + \frac{\partial x_2}{\partial X_2}\cdot\frac{\partial x_2}{\partial X_2}a_{22}
$$

where A_{11} etc. are the components of a in the new coordinate system X.

Author Index

Subject Index

RETURN TO: **CHEMISTRY LIBRARY**

100 Hildebrand Hall • 510-642-3753

LOAN PERIOD 1	2	3
4	1-MONTH USE	6

ALL BOOKS MAY BE RECALLED AFTER 7 DAYS.

Renewals may be requested by phone or, using GLADIS,
type **inv** followed by your patron ID number.

DUE AS STAMPED BELOW.

MAY 24 2003

FORM NO. DD 10   UNIVERSITY OF CALIFORNIA, BERKELEY
2M 5-01   Berkeley, California 94720–6000